# The Power of Interaction

'

## ACM Distinguished Dissertations

**1982**
*Abstraction Mechanisms and Language Design,* Paul N. Hilfinger
*Formal Specification of Interactive Graphics Programming Language,* William R. Mallgren
*Algorithmic Program Debugging,* Ehud Y. Shapiro

**1983**
*The Measurement of Visual Motion,* Ellen Catherine Hildreth
*Synthesis of Digital Designs from Recursion Equations,* Steven D. Johnson

**1984**
*Analytic Methods in the Analysis and Design of Number-Theoretic Algorithms,* Eric Bach
*Model-Based Image Matching Using Location,* Henry S. Baird
*A Geometric Investigation of Reach,* James U. Korein

**1985**
*Two Issues in Public-Key Cryptography,* Ben-Zion Chor
*The Connection Machine,* W. Daniel Hillis

**1986**
*All the Right Moves: A VLSI Architecture for Chess,* Carl Ebeling
*The Design and Evaluation of a High Performance Smalltalk System,* David Michael Ungar

**1987**
*Algorithm Animation,* Marc H. Brown
*The Rapid Evaluation of Potential Fields in Particle Systems,* Leslie Greengard

**1988**
*Computational Models of Games,* Anne Condon
*Trace Theory for Automatic Hierarchical Verification of Speed-Independent Circuits,* David L. Dill

**1989**
*The Computational Complexity of Machine Learning,* Michael J. Kearns
*Uses of Randomness in Algorithms and Protocols,* Joe Kilian

**1990**
*Using Hard Problems to Create Pseudorandom Generators,* Noam Nisan

**1991**
*Performance Analysis of Data Sharing Environments,* Asit Dan
*Redundant Disk Arrays: Reliable, Parallel Secondary Storage,* Garth A. Gibson
*The Power of Interaction,* Carsten Lund

# The Power of Interaction

Carsten Lund

The MIT Press
Cambridge, Massachusetts
London, England

This book was printed and bound in the United States of America.

Library of Congress Cataloging-in-Publication Data

Lund, Carsten, 1963–
    The power of interaction / Carsten Lund.
        p.    cm. — (ACM distinguished dissertations)
    Includes bibliographical references (p.      ) and index.
    ISBN 0-262-12170-0
    1. Proof theory. I. Title. II. Series.
QA9.54.L86 1992
511.3—dc20                                               92-18380
                                                              CIP

*Dedicated to Linda M. Sellie*

# Contents

# Preface and Acknowledgements

The theory of computation is the study of the inherent difficulty of computational problems. The study of efficient proof systems has been central in the theory of computation. The most important discovery in the theory of computation is that the class of problems that have efficient proof systems – *NP* – contains problems that are the hardest – *NP*-complete – in the class, e.g., if one of the complete problems can be solved efficiently, then every problem in the class can be solved efficiently.

In 1985, *interactive proof systems* were put forth as a randomized analog of traditional proof systems. Some problems were shown to have randomized proof systems in this new model that most likely do not have traditional proof systems, but it was widely believed that these randomized interactive proof systems did not have significantly more power than traditional *NP* proof systems.

This book describes a new technique for constructing randomized proofs and shows that interactive proof systems are far more powerful than traditional *NP* proof systems.

This book contains the first in a series of results that have gotten attention even in the popular press. The techniques described here are also the basis for the subsequent results and therefore this book can be seen as an introduction to these new developments.

This book is a revision of my doctoral dissertation, which was completed in January 1991 at the University of Chicago. A general background in theoretical computer science on the level of a second year graduate student is assumed of the reader but no prior knowledge of the theory of computation is assumed.

**Acknowledgements.** First I would like to thank my advisors: Joan Boyar, Janos Simon, László Babai and Lance Fortnow. They have all greatly influenced my research at the University of Chicago.

I spent three wonderful years in Chicago, for a large part due to the great atmosphere in the Computer Science Department. There is a good blend of social and research activities that makes it a pleasure to work there. I would like to thank the graduate students for their great friendship. I would like to thank my officemate Sundar Vishwanathan for many good discussions and proof reading.

The research described in this book is joint work with a number of outstanding researchers and I would like to thank each for them: László Babai, Lance Fortnow, Howard Karloff and Noam Nisan.

I was during the first two and a half years supported partly by a fellowship from the University of Århus, Denmark. I am very thankful for this generous opportunity.

Thanks to AT&T Bell Laboratories for their generous support during my revision of this book.

Carsten Lund
Murray Hill, New Jersey
June 1992

# The Power of Interaction

# 1

## Introduction

Under the siege of Troy the princess Helen was spending her spare time talking to the Hellenic gods. Trying to keep her mind off her terrible situation, Helen asked the gods to prove theorems to her. The gods, themselves getting tired of the prolonged war, agreed. In this book we are going to study which theorems the gods were able to convince Helen about.

Theorems have been proven since the time of the first mathematicians.

*But what is a proof?*

In a math textbook a proof consists of a sequence of letters and symbols. That the sequence is located in a math book does not make it a proof. What makes it a proof is that people who read the proof can be convinced that the proof is valid, and thus that the theorem is true.

The French Mathematician Fermat (1601-1665) wrote in the margin of one of his papers that he had a proof of

$$\forall n > 2 : \forall a, b, c : a^n + b^n \neq c^n,$$

but that the margin did not provide enough space for the proof. We may strongly believe in Fermat's honesty and mathematical ability but we will not consider the proof he had in mind convincing, since there has been no verification of his proof. Hence a *proof system* involves a proving part and a verifying part.

The verifying part checks the validity of the proof and is not supposed to have any special insight in the problem, whereas the proving part needs to have enough deep insight in the problem to be able to give a convincing proof. In other words Helen is human and therefore does not know how to prove the divine theorems. She only has to verify the gods' proofs.

Since Plato, logicians have studied reasoning about formal systems. Traditionally they have considered proofs as a sequence of logic expressions, where one expression follows the next, using simple rules. To verify a proof, the verifier only has to check every step of the sequence, thus making the verification simple.

Since Helen is talking with the gods, she can ask questions and get answers. Helen has to be aware that the gods have greater abilities than herself, and they could use their cunning skills to predict her future questions. So to make sure that they can not predict her question, she tosses a coin and lets the coin flip determine her next question. Not even the gods are capable of predicting her future coin flips.

### When is a proof valid?

If the gods try to predict Helen's future random coin flips and use this prediction when making their proof, they may be able to make Helen accept a false theorem. If the gods' prediction by chance turns out to be right, this strategy can convince Helen about false theorems.

In logic, a proof is either valid or invalid, whereas in mathematics some proofs are so complicated that they are not totally convincing. There may still be a small "chance" that the proof may not be valid. For example the proof of the Classification of Finite Simple Groups has 5000 pages written over a period of 100 years. So in traditional mathematics there is the possibility of proofs turning out to be invalid. In this light it is not so big of a conceptual jump to allow Helen to make infrequent mistakes. Say that she in every $10^{123}$ proofs makes one mistake, where $10^{123}$ is the number of atoms in the universe.

### When is a proof efficient?

Since Helen is human, she wants to be able to verify a proof in her lifetime. Hence a proof has to be short.

One thing traditionally not considered in logic is the length of the proof sequence, which could be extremely long, and thereby the verification cannot be done efficiently even if each single step is easy. We are going to take this difficulty into account and restrict the length of the gods' proofs.

In mathematics clearly a proof has to be efficient since mathematics involves human beings.

The study in logic of "efficient" proof systems is young. It is only recently that the concept of efficiency has been formalized in the theory of computation. Efficient proof systems were introduced in work by Cook, Karp and Levin in the early 1970's and have had a great significance for the whole theory of computation.

In 1985 Goldwasser, Micali and Rackoff [40] and independently Babai [5] introduced *interactive proof systems*. This model describes the situation where Helen is interacting with *one* god. Subsequently Goldwasser, Micali and Wigderson [39] gave an interactive proof for a problem which is not believed to have a short traditional proof. But until the work described in this book it was an open problem as to how many problems have interactive proofs.

In 1988 Ben-Or, Goldwasser, Kilian and Wigderson [14] introduced *multiple prover interactive proof*. This model describes the situation in Troy, where Helen is talking to two gods that are on opposite sides in the war and they are therefore not in the same room. Helen therefore has to walk from room to room and talk with them separately. Hence each god will be able neither to communicate with the other god nor hear Helen's question to the other god. Thus Helen can try to play one god against the other and thereby learn even more theorems.

We show that Helen indeed can verify theorems that she herself could not prove. Our results shows that interaction with two provers/gods and randomization can give exponential speedup in efficiency compared to traditional proof systems. Furthermore we show that even with only one prover/god Helen can verify theorems that are believed to be hard.

# 2

# Preliminaries

## 2.1 Basic Definitions

### 2.1.1 Notation and Basics

Given functions $f, g : \mathbf{N} \to \mathbf{N}$, where $\mathbf{N}$ is the natural numbers, we say that $f = O(g)$ if there exist constants $c, N$ such that $\forall n \geq N : f(n) \leq cg(n)$. Conversely $f = \Omega(g)$ if and only if $g = O(f)$.

A *string* over some alphabet $\Sigma$ is a finite sequence of elements from $\Sigma$. The *length* is the number of elements in the sequence. $\Sigma^*$ denotes the set of all strings over alphabet $\Sigma$. We restrict, without loss of generality, our attention to the case where $\Sigma = \{0, 1\}$. A *language* is a subset of $\Sigma^*$.

### 2.1.2 Boolean Formulas

A *Boolean function* is a function from $\{0, 1\}^n$ into $\{0, 1\}$, where $n$ is a positive integer. There are clearly $2^{2^n}$ such functions. In the study of Boolean functions there are some functions that are simpler than others in that they have a short description. We define two common ways of describing Boolean functions.

A *Boolean formula* is a labelled tree. A leaf is labelled by a Boolean constant 0 or 1 or by a *literal* ($x_i$ or its negation $\overline{x}_i$). The internal nodes are

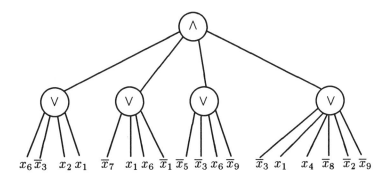

Figure 2.1: A CNF formula.

labelled by Boolean operations ∨ or ∧. Internal nodes labelled ∨ are called OR gates and internal nodes labelled ∧ are called AND gates. The members of the set $\{x_1, x_2, \ldots, x_n\}$ are called the variables. A *truth assignment* is a function from the variables to the set of Boolean values $\{0, 1\}$. A truth assignment $v$ to the variables induces an assignment of a Boolean value to each gate in the formula. If every descendant of an OR gate is assigned 0 then the gate is assigned 0 otherwise the gate is assigned 1. Similarly if all the descendants of an AND gate are assigned 1 then the gate is assigned 1 otherwise 0. A formula is *satisfied* if and only if the root is assigned the value 1.

The *size* of a formula is the number of nodes in the tree. The *depth* is the length of the longest path from the root to a leaf.

A Boolean formula in conjunctive normal form (CNF) is a Boolean formula of depth 2 where the root is an AND gate and the nodes on the next level are OR gates. See Figure 2.1.

A 3-CNF formula is a CNF formula where each OR consists of 3 literals.

The *Boolean circuit* is the other form of description of Boolean functions that we will use. A Boolean circuit is an acyclic directed labelled graph with indegrees 0 and 2. There is a special *output* node that has outdegree 0. All other nodes have outdegree at least 1. A node with indegree zero corresponds to the leaves of the Boolean formula. It is labelled by either a Boolean constant or a literal. The other nodes correspond to the internal node in the formulas and are labelled with a Boolean operation $\vee$ or $\wedge$. Since the graph is acyclic each node, just like each node in a formula, computes a Boolean function. The function computed by the circuit is the one computed at the output node.

## 2.1.3   Arithmetic Formulas and Expressions

The main techniques of this book use basic algebra such as polynomials and roots of polynomials.

We assume that basic concepts such as rings, fields, polynomials over a ring and roots of a polynomial are known to the reader. We are using *prime fields*, the ring of integers $\mathbf{Z}$ and the field of rationals $\mathbf{Q}$. The prime field $\mathbf{F}_p$ denotes the integers modulo a prime $p$.

We will extend Boolean functions to functions defined over rings or fields. We will always let the Boolean values 0 and 1 correspond to the integers 0 and 1 in the algebraic setting. We say that a function $g : R^n \to R$ *interpolates* a Boolean function $f : \{0,1\}^n \to \{0,1\}$ if for every $\alpha \in \{0,1\}^n$,

$$g(\alpha) = f(\alpha).$$

The following Proposition states that we always can find such a $g$ that interpolates any given $f$.

**Proposition 2.1.1** *Given a Boolean function* $f : \{0,1\}^n \to \{0,1\}$ *there exists a unique multilinear function* $g : \mathbf{Z}^n \to \mathbf{Z}$ *interpolating* $f$.

**Proof:**   Let us first look at the special case where $f$ is an indicator function. Let us define $\delta_\alpha$ for $\alpha, \beta \in \{0,1\}^n$ by

$$\delta_\alpha(\beta) := \begin{cases} 1 & \text{if } \beta = \alpha \\ 0 & \text{otherwise.} \end{cases}$$

These functions can be interpolated by the following multilinear polynomial

$$p_\alpha(x_1, x_2, \ldots, x_n) := \prod_{i=1}^{n}(1 - \alpha_i - x_i + 2x_i\alpha_i),$$

where $\alpha = (\alpha_1, \alpha_2, \ldots, \alpha_n)$.

Hence, if we define

$$g(x) := \sum_{\alpha \in \{0,1\}^n} f(\alpha) p_\alpha(x)$$

then we get that $g$ is a multilinear polynomial interpolating $f$.

The uniqueness follows from induction in the number of variables. Observe that it is enough to show that any multilinear polynomial that interpolates the Boolean constant zero function must be the zero polynomial.

For $n = 1$ this follows since any non-zero linear integer polynomial has at most one integer root.

For $n > 1$ let $p$ be a multilinear polynomial that interpolates the Boolean constant zero function. Since $p(x_1, x_2, \ldots, x_n)$ is a multilinear polynomial it can be written as $x_n p'(x_1, x_2, \ldots, x_{n-1}) + p''(x_1, x_2, \ldots, x_{n-1})$, where $p', p''$ are multilinear polynomials. But by letting $x_n = 0$ we know that $p''$ interpolates the Boolean constant zero function on $n - 1$ variables and hence by the inductive hypothesis it is the all zero polynomial. Hence, similarly we get that $p'$ is the zero polynomial and, therefore, we get that $p$ is the zero polynomial. $\blacksquare$

One problem with the above Proposition is that it does not say that if $f$ is simple to describe, then $g$ is simple to describe. Or if a simple description of $g$ does exist then it may be hard to go from a simple description of $f$ to a simple description of $g$. For example, if $f$ is described by a small formula and all assignments satisfy the formula, then $g$ is the polynomial 1. But to determine if $g = 1$ is *coNP*-complete (see the definition of *coNP* and completeness later) and is therefore likely to be hard to decide. So we show a way of describing arithmetic functions, which can then be used to give an effective way of going from descriptions of Boolean functions to descriptions of interpolating arithmetic functions.

An *arithmetic formula* is a labelled binary tree. The leaves are labelled by the constants 0, 1 and −1 or by a variable from $\{x_1, x_2, \ldots, x_n\}$. The internal nodes are labelled by a binary *arithmetic operation* + or ·. See Figure 2.2.

An arithmetic formula $F$ naturally associates a polynomial $p_v$ with each node $v$. Note that we only allow the constants 0, 1 and −1 in a formula, hence

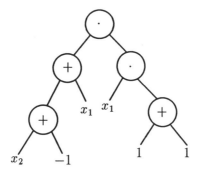

Figure 2.2: An arithmetic formula.

the polynomials are independent of the ring $R$, since 0, 1 and $-1$ belong to every ring $R$. The *degree* of an arithmetic formula is the maximal degree of any of the $p_v$'s. Let $p_F$ be the polynomial associated with the root. Note that the degree of a formula is at most the size of the formula, and that $p_F$ can be evaluated at a point $(\alpha_1, \alpha_2, \ldots, \alpha_n) \in R^n$ by using, at most, $size(F)$ arithmetic operations.

**Example 2.1.2** *Let $F$ be the formula from Figure 2.2. Then $p_F$ is equal to* $((x_2 + (-1)) + x_1)(x_1(1+1)) = 2x_1 x_2 - 2x_1 + 2x_1^2$.

The *arithmetic expression* is a generalization of the arithmetic formula. Whereas formulas are easy to evaluate, expressions can be hard to evaluate. Formulas can be evaluated by a linear number of arithmetic operations, whereas the expressions, as we shall see later, can be much harder to evaluate.

An arithmetic expression is a labelled tree. The leaves are labelled by a constant 0, 1, $-1$ or by a variable from $\{x_1, x_2, \ldots, x_n\}$. The internal nodes have either one or two descendants. The nodes with two descendants are labelled by an arithmetic operation $+$, or $\cdot$, whereas the node with one descendent is labelled by a *arithmetic operator* $\sum_{x_i=0}^{1}, \prod_{x_i=0}^{1}, \coprod_{x_i=0}^{1}$ or $\mathcal{R}x_i$. These operators take a polynomial and transform it into another polynomial. For example $\sum_{x_1=0}^{1}(x_1 + x_2) = 1 + 2x_2$. The operators $\sum$ and $\prod$ have the normal mathematical meaning, whereas $\coprod$ and $\mathcal{R}x_i$ are special and are introduced to be used in the proofs of this book later. Formally the operators are defined as

- $\sum_{x_i=0}^{1} f(x_1, .., x_i, .., x_n) := f(x_1, .., 0, .., x_n) + f(x_1, .., 1, .., x_n)$.

- $\prod_{x_i=0}^{1} f(x_1, .., x_i, .., x_n) := f(x_1, .., 0, .., x_n) f(x_1, .., 1, .., x_n)$.

- $\coprod_{x_i=0}^{1} f(x_1, x_2, \ldots, x_n) := 1 - \prod_{x_i=0}^{1}(1 - f(x_1, x_2, \ldots, x_n))$.

- $\mathcal{R}x_i f(x_1, x_2, \ldots, x_n) := f(x_1, x_2, \ldots, x_n) \pmod{x_i^2 - x_i}$.

Just as for arithmetic formulas, we can associate a polynomial $p_E$ with every expression $E$. But contrary to formulas the degree of $p_E$ can be exponential in the size of $E$.

A variable $x_i$ is called a *free variable* if there exists a path from the root to a leaf, that does not contain a node labelled by any of the operators $\sum_{x_i=0}^{1}, \prod_{x_i=0}^{1}$ or $\coprod_{x_i=0}^{1}$. The reason for this is that the scope of $x_i$ ends with those operators. For example $\sum_{x_1=0}^{1} \sum_{x_2=0}^{1} \cdots \sum_{x_n=0}^{1} F(x_1, x_2, \ldots, x_n)$, where F is an arithmetic formula, has no free variables.

If an expression $E$ does not have any free variables then $p_E$ is just a constant in $R$ and we call it the *value* of $E$ in $R$. Note that the absolute value of the value of an expression in the integers can be as big as doubly exponential. For example $\prod_{x_1=0}^{1} \prod_{x_2=0}^{1} \cdots \prod_{x_n=0}^{1}(1 + 1)$ has value $2^{2^n}$ in $\mathbf{Z}$.

# 2.2 Computational Models

This section defines different mathematical models of computation. The computational devices in each model takes some input string $x$ over an alphabet $\Sigma$ and then, after some computation, either accepts or rejects $x$.

We start out by defining a model that represents some idealized version of the present day computer. Later we will define models that add features to the basic model.

## 2.2.1 Deterministic Computation

Our basic computational device is the Deterministic $k$-tape Turing machine (TM), where $k$ is some integer constant. TMs were introduced by Turing [75]

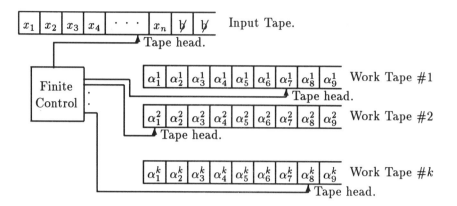

Figure 2.3: A $k$-tape Turing Machine.

in 1936. It models the everyday computer and operates, like the computer, in discrete time steps. At each time step it changes its internal configuration according to some simple rules. A TM consists of *tapes*, *tapeheads* and a *finite control*. See Figure 2.3.

- Tapes are infinite sequences of *cells*. Each cell contains one symbol from a finite alphabet $\Delta$ containing $\Sigma$. The $k$-tape TM has $k$ *work tapes* and one *input tape*. The work tapes are used to store information during the computation. Initially all cells on the work tapes contain a special character ⊬ called the *blank* character. The input tape contains the input string, which is stored in the first $n$ cells, where $n$ is the length of the input string. The rest of the input tape contains blanks.

- Each tape has a tapehead that *scans* one tape cell on that tape. Initially each head scans the first cells on each tape.

- The finite control is in a *state*, which is an element from a finite set $Q$. Initially it is in the *start state* $q_0 \in Q$.

From one time step to the next a TM does the following.

- It changes the symbol on each work tape cell being scanned.

- It moves the heads one cell either to the left (L), to the right (R) or it stays (S) at the same cell.

- It changes the state of the finite control.

Or

- It chooses to halt the computation and accepts (respectively rejects) by entering a special accepting state $q_A \in Q$ (respectively a special rejecting state $q_R \in Q$).

The action performed is determined by the *transition diagram* $\delta$, which is a function from the symbols being scanned and the state of the finite control to an action. Formally $\delta : Q \times \Delta^{k+1} \to Q \times \Delta^k \times \{L, R, S\}^{k+1}$.

An *instantaneous description* (ID) of a TM is a description of the configuration of the machine at some instant during the computation. The configuration consists of the contents of the tapes, the locations of the heads and the state of the finite control. Observe that we only need to describe a finite prefix of each tape since we can assume that the rest of the tape consists of cells with the blank symbol. Formally an ID is a tuple $(q, x, \alpha^1, \alpha^2, \ldots, \alpha^k, h, h_1, h_2, \ldots, , h_k)$ of a state, the input string, strings that describe the work tapes, the position of the input head and the positions of the work heads. For example the ID corresponding to the start configuration is $(q_0, x, \epsilon, \epsilon, \ldots, \epsilon, 1, 1, \ldots, 1)$, where $\epsilon$ is the empty string.

During the computation of a TM $M$ on input $x$, one of three situations arise:

1. $M$ halts and accepts after a finite amount of time.

2. $M$ halts and rejects after a finite amount of time.

3. $M$ does not halt but keeps on running indefinitely.

We say that $M$ *accepts* $x$ if the first situation arises else we say that $M$ *rejects* $x$. The language *recognized* by $M$ is

$$L(M) := \{x \mid M \text{ accepts } x\}.$$

We will in the rest of this book only discuss TMs that always halt. This is not a limitation on the results, since we are looking only at machines that can always be made to halt on all inputs.

We extend the definition of TM to allow the TM to compute a function of the input instead of just recognizing a language. Here the TM has an extra tape called the *output tape*. It is a *one-way* (the tape head can only move to the right) tape onto which the machine can write only the symbols 0 and 1. During the computation the machine writes a string $y \in \{0,1\}^*$ on the output tape before it halts. This string is called the output of the TM. This implies that a TM defines a function from $\{0,1\}^*$ into $\{0,1\}^*$. Observe that the function is defined for all inputs, since we require that the TM halts on all inputs.

### 2.2.2   Probabilistic Computation

One generalization of TM called the probabilistic Turing machine (PTM) is to allow the machine to flip coins during the computation. This is done by letting the PTM have a *random tape*, which is *read-only* (the symbols can not be changed) and one-way. Each cell of the random tape computation contains independently of the other cells 0 or 1 with probability $\frac{1}{2}$. The computation of the machine depends, therefore, on the content of the random tape. Thus on input $x$ the PTM accepts sometimes and rejects at other times. We assume that the machine always halts, no matter what the content of the random tape. For each input $x$, there is an upper bound $u_x$ on how many cells on the random tape the machine reads. A PTM $M$ accepts an input $x$ with probability $p$ if

$$|\{r \in \{0,1\}^{u_x} \mid M \text{ accepts } x \text{ with } r \text{ on the random tape}\}| = p2^{u_x}.$$

$M$ recognizes a language $L$ if and only if for $x \in L$, $M$ accepts with probability at least $\frac{2}{3}$ and if $x \notin L$ then $M$ rejects with probability at least $\frac{2}{3}$.

### 2.2.3   Non-Deterministic Computation

One way to look at deterministic computation is that the TM $M$ constructs a proof of the fact, that the input belongs to $L(M)$. The proof can be written as a sequence of IDs that show the configuration of $M$ at each time step. It is easy to check whether a proof is valid just by looking at the sequence. This raises the question is it easier to check a proof than to construct a proof? To study this question we define machines that only have to check proofs. The proofs, if they exists, are magically given to the machines.

Non-deterministic computation is defined in terms of non-deterministic Turing machines (NTM). The only difference between TM and NTM is that the transition diagram is no longer just a function of the configuration but the NTM can go to one or more new configurations. Formally, $\delta$ is a relation:

$$\delta \subset \left(Q \times \Delta^{k+1}\right) \times \left(Q \times \Delta^k \times \{L, R, S\}^{k+1}\right).$$

If a NTM $M$ is in a state $q$ and scans symbols $\alpha \in \Delta^{k+1}$ then it can enter state $q'$, change the work symbols to $\alpha' \in \Delta^k$ and move the head accordingly to $\beta \in \{L, R, S\}^{k+1}$ if and only if $((q, \alpha), (q', \alpha', \beta)) \in \delta$.

The computation of a NTM is a tree (*the computation tree*) labelled with IDs, where the ID at the root correspond to the NTM at the start of the computation. The descendants of a node with label $I$ are labelled with the different IDs that can follow $I$ in the computation in one step. The leaves of the tree are labelled with accepting and rejecting IDs. A path from the root to a leaf is called a *computation path*. All paths will be finite since we assume that the NTM halts for all inputs and for all computation paths.

We say that a NTM *accepts* an input if there exists an accepting computation path on that input, *i.e.,* a path from the root of the computation tree to an accepting leaf. One way to think about this is that the NTM always, out of a set of possible actions, magically chooses an action that leads to an accepting configuration, if such a choice exists.

Observe that the computation tree defines a Boolean formula with all internal nodes being OR gates, accepting leaves being true, rejecting leaves being false and the formula being true if and only if $M$ accepts the input.

## 2.2.4   Alternating Computations

One generalization of NTMs due to Chandra, Kozen and Stockmeyer [20] is to introduce AND gates into the formula defined by the computation tree.

This is formalized by partitioning the states of the alternating Turing machine (ATM) into two sets: the *existential* states and the *universal* states. Just as for the NTM, the ATM can go from one configuration to one or more configurations determined by the transition diagram. In the formula defined by the computation tree, the nodes that are in an existential state are OR gates and nodes that are in a universal state are AND gates. An ATM accepts an input if and only if the formula is satisfied.

## 2.2.5   Interactive Proof Systems

Interactive proof systems are a generalization of NTMs that incorporates randomness. This model was introduced by Goldwasser, Micali and Rackoff [40] in 1985 and independently by Babai [5] also in 1985. Interactive proof systems can be modeled as a game between 2 players: a *verifier* and a *prover*. The verifier's computation is restricted in some way (See section 2.3.5), whereas the prover has no restriction on its computational power. Given a language $L$ and a string $x$. It is the verifier's objective to figure out if $x$ is a member of $L$. The prover's objective is to make the verifier accept $x$ as a member of $L$. So if $x \in L$ then the prover tries to help the verifier, but when $x \notin L$ then the prover tries to cheat the verifier into accepting $x$. The verifier can ask questions of the prover about $x$ but the verifier can not trust the answers.

Formally, an interactive proof system is a pair (P,V) of finite control machines with their own work tapes, an input tape that they both can read, a *communication tape* that P writes on and V reads, a communication tape that V writes on and P reads and a random tape that V reads. See Figure 2.4. Both the communication and random tapes are one-way, hence the heads on those tapes can only move to the right.

The computation in this model proceeds in *rounds*. One round consists of some local computation by the verifier during which the verifier asks the prover a question by writing a message on the communication tape. Thereafter the prover does some local communication during which it reads the question and writes an answer. The verifier can, during its computation, decide to halt the computation and accept or reject. An interactive proof system $(P, V)$ *recognizes* a language $L$ if and only if

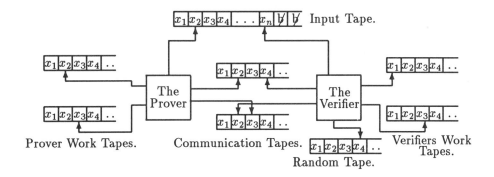

Figure 2.4: An interactive proof system.

- for strings in $L$ there is a large chance that the verifier accepts.

- for strings not in $L$ then for all cheating provers there is only a small chance that the verifier accepts.

The first property is that the proof system is *complete* and the second is that it is *sound*. For soundness we do not even restrict our cheating provers to be Turing machines. They may even compute any non-recursive function.

Formally we say:

**Definition 2.2.1** *$L$ is recognized by the interactive proof system $(P, V)$ if*

- $\forall x \in L : Pr_R[V$ *accepts when communicating with* $P] \geq \frac{2}{3}$.

- $\forall x \notin L : \forall P' : Pr_R[V$ *accepts when communicating with* $P'] \leq \frac{1}{3}$.

Observe that the prover can not see the random tape of the verifier. Babai in [5] independently of Goldwasser, Micali and Rackoff [40] introduced similar games (Arthur-Merlin games), where the prover could see the random bits after the verifier has read them. We call these *public-coin* interactive proof systems.

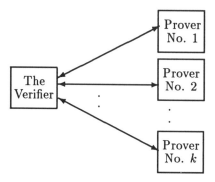

Figure 2.5: A multiple-prover interactive proof system.

Observe that NTMs are special cases of of interactive proof systems since the prover can lead the verifier down an accepting path in the computation tree. Also PTMs are special cases since the verifier does not have to use the prover.

## 2.2.6   Multiple Prover Interactive Proof Systems

Multiple Prover Interactive Proof Systems (MIPS) is a further extension of proofs systems introduced by Ben-Or, Goldwasser, Kilian and Wigderson in [14]. In MIPS the verifier can talk to more than one prover (See Figure 2.5.) The provers can, before the computation, decide a strategy about how to cheat the verifier. But once the computation begins, the provers can not communicate between themselves, or see the conversation between the other provers and the verifier.

Figure 2.6 give the obvious relationships between the computational models that we have introduced.

## 2.2.7   Computation Relative to an Oracle

In the study of computation, it has been important to consider computations that get some advice on how to proceed.

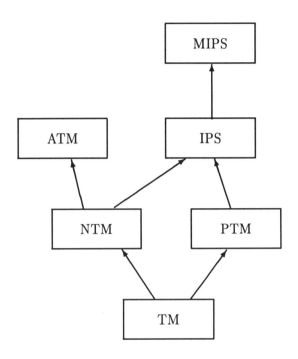

Figure 2.6: The relationship between computation models.

Baker, Gill and Solovay [11] formalized this by considering the model of oracle Turing machines (OTM). An OTM is a Turing machine that has an additional tape, called the *oracle tape* and 3 special states $(q_?, q_0, q_1)$. The computation of an OTM $M^A$ depends on oracle $A \subset \Sigma^*$. $M^A$ can, during its computation, write down a string $y$ on the oracle tape and then enter the special state $q_?$. From that state, $M^A$ magically enters $q_1$ if $y$ belongs to the oracle $A$ otherwise $M^A$ enters the state $q_0$. Hence, an OTM can get answers about membership in the oracle in one step. We say that the computation is performed *relative* to the oracle $A$.

Similarly we have non-deterministic (respectively probabilistic or alternating) oracle Turing machines by letting the Turing machine be non-deterministic (respectively probabilistic or alternating). By letting the verifier have access to an oracle we, similarly, get interactive proof systems relative to oracles.

# 2.3   Complexity Classes

Complexity theory looks at the amount of resources it takes machines to solve problems in the different models. The resources we are interested in are *time* and *space*. In this section we define classes of language that are recognizable by different computational models when these resources are restricted.

## 2.3.1   Deterministic Complexity Classes

The definitions of time and space for deterministic TM model the concepts of running time and random access memory for everyday computer programs. Let $M$ be a deterministic TM. The time $Time_M(x)$ that $M$ uses on input $x$ is just the number of steps $M$ takes on input $x$ before either accepting or rejecting. The space $Space_M(x)$ is the number of work tape cells it uses. Observe that the input tape cells are not counted as space. This allows us to look at space bounds of order less than the input size.

We are interested in the time taken and space used by the TM as functions of the size of the input. So given $s, t : \mathbf{N} \to \mathbf{N}$ we say that a TM $M$ *works* in time $t$ and space $s$ if and only if

$$\forall x \in \Sigma^* : Time_M(x) \leq t(|x|) \text{ and } Space_M(x) \leq s(|x|).$$

Denote the collection of TMs that work in time $O(t)$ (respectively space $O(s)$) by $\mathcal{T}_t$ (respectively $\mathcal{S}_s$). We use $DTIME(t)$ (respectively $DSPACE(s)$) for languages that are recognized by a TM from $\mathcal{T}_t$ (respectively $\mathcal{S}_s$). So

$$DTIME(t) := \{L \subset \Sigma^* \mid L = L(M) \text{ for some TM } M \in \mathcal{T}_t\}$$

$$DSPACE(s) := \{L \subset \Sigma^* \mid L = L(M) \text{ for some TM } M \in \mathcal{S}_s\}.$$

Sometimes we are interested in bounding both the resources at the same time. Bruss and Meyer [18] defined the class $DTISP(t, s)$ as the class of languages recognized in deterministic time $t$ and space $s$.

$$DTISP(t, s) := \{L \subset \Sigma^* \mid L = L(M) \text{ for some TM } M \in \mathcal{T}_t \cap \mathcal{S}_s\}.$$

We restrict our attention to "nice" resource bounds. By nice we mean *fully time constructible* and/or *fully space constructible*. A function $t$ is fully time constructible if there exist a TM $M$ such that on all inputs of length $n$, $M$ computes $t(n)$ in binary in time $O(t(n))$. Similarly a function $s$ is fully space constructible if there exist a TM $M$ such that $M$ on all inputs of length $n$ computes $1^{s(n)}$ in space $O(s(n))$. Most "natural" functions are nice in the sense of being fully time/space constructible and we restrict our time (respectively space) bounds to be fully time (respectively space) constructible. Examples of nice functions are $\lfloor \log n \rfloor, \lceil \log^3 n \rceil, n, n^{24}, \lceil 2^{\sqrt{n}} \rceil$. We are interested in classes where we simultaneously restrict time and space. So we define that a pair of function $(t, s)$ are *fully time-space constructible* if there exist a TM $M$ such that on all inputs of length $n$, $M$ computes $t(n)$ in binary and $1^{s(n)}$ using $O(t(n))$ time and $O(s(n))$ space.

It is easily observed (see [45, Theorem 12.10a]) that a TM that uses at most $t$ time uses at most $t$ space. So

$$DTIME(t) = DTISP(t, t).$$

For space it is observed (see [45, Theorem 12.10b]) that if a TM $M$ uses at most $s$ space then there exists a constant $c$ such that $M$ uses at most $c^s$ time, because we have assumed that $M$ always halts. Hence

$$DSPACE(s) = \bigcup_{c=1}^{\infty} DTISP(c^s, s).$$

We define the class $P$ of languages recognizable in polynomial time that traditionally has been accepted as the class of problem with efficient solutions.

$$P := \bigcup_{k=0}^{\infty} DTIME(n^k).$$

We call language in $P$ *tractable* and languages outside $P$ *intractable*. A subset of $P$ that uses a small amount of space is

$$L := DSPACE(\log n)).$$

We define the class *PSPACE* as the set of languages that have algorithms that are efficient with respect to the amount of space they use.

$$PSPACE := \bigcup_{k=0}^{\infty} DSPACE(n^k).$$

A class that is known to contain language not in $P$ is

$$EXP := \bigcup_{k=0}^{\infty} DTIME(2^{n^k}).$$

We define classes of functions in a similar way. $fP$ is the class of functions computable in deterministic polynomial time. In general $fC$ denotes a class of functions, where $C$ is the corresponding language class.

## 2.3.2   Probabilistic Complexity Classes

Time is defined in probabilistic computation as the length of the longest computation path. In other words, it is the height of the computation tree. Space is defined as the maximum number of work cells used on any computation path.

The definitions of $BPTIME(t)$, $BPSPACE(s)$ and $BPTISP(t,s)$ are analogous to the deterministic case.

The class $BPP$ is defined to be the class of languages that are recognizable in polynomial time by a probabilistic Turing machine.

$$BPP := \bigcup_{k=0}^{\infty} BPTIME(n^k).$$

## 2.3.3   Non-Deterministic Complexity Classes

Time and space is defined as for probabilistic computations and the definition of $NTIME(t)$, $NSPACE(s)$ and $NTISP(t,s)$ are similar to the deterministic and probabilistic cases.

The class *NP* is defined to be the class of languages that are recognizable in polynomial time by a nondeterministic Turing machine. Another way to say this is that *NP* is the class of languages that are verifiable in deterministic polynomial time. Hence, given a computation path, it is possible to verify in polynomial time that the path accepts the input.

$$NP := \bigcup_{k=0}^{\infty} NTIME(n^k).$$

The most famous open problem in complexity theory and maybe in all of computer science is the question whether *P* equals *NP*. It is widely believed in the scientific community that $P \neq NP$. But a solution, one way or the other, seems to be far away from present day proof techniques. A problem related to $P \overset{?}{=} NP$ is whether *NP* is closed under complement. Generally given a class of languages $\mathcal{C}$ we define the complement to $\mathcal{C}$ as $co\mathcal{C} := \{\overline{L} \mid L \in \mathcal{C}\}$, where $\overline{L} = \{x \mid x \notin L\}$. If $NP \neq coNP$ then it would imply that $P \neq NP$, since $P = coP$.

We also define *NP*'s polynomial space and exponential time equivalents

$$NPSPACE := \bigcup_{k=0}^{\infty} NSPACE(n^k)$$

$$NEXP := \bigcup_{k=0}^{\infty} NTIME(2^{n^k}).$$

There are two important results about nondeterministic space. The first relates nondeterministic space to deterministic space:

**Fact 2.3.1 (Savitch [64])** *If s is fully space-constructible and $s(n) \geq \log n$ then*

$$DSPACE(s) \subset NSPACE(s) \subset DSPACE(s^2)$$

*and therefore*

$$PSPACE = NSPACE.$$

The second states that nondeterministic space is closed under complement.

**Fact 2.3.2 (Immerman and Szelepcsényi [47, 73])** *Let $s(n) \geq \log n$ then*

$$co\,NSPACE(s) = NSPACE(s).$$

## 2.3.4 Alternating Complexity Classes

We define time and space of an ATM just as for a NTM and we get the classes $ATIME(t)$, $ASPACE(s)$ and $ATISP(t,s)$. Similarly we get the classes $AP$, $APSPACE$ and so on. Some relations between alternative classes and deterministic classes are known.

**Fact 2.3.3 (Chandra, Kozen and Stockmeyer [20])**

- *For $s(n) \geq \log n$, $ASPACE(s(n)) = \bigcup_{c>0} DTIME(c^{s(n)})$.*
- *For $t(n) \geq n$, $ATIME(t(n)) \subseteq DSPACE(t(n)) \subseteq ATIME(t(n)^2)$.*

*and therefore*
$$AP = PSPACE \text{ and } APSPACE = EXP.$$

Ruzzo [63] studied the class *NC*, which is the class of languages that have polynomial size and poly-logarithmic depth uniform circuits (see section 2.3.6). He looked at alternating machines that are allowed only poly-logarithmic time and logarithmic space. To make the computations depend on all the inputs he introduced a variant of the ATM. In this variant the input is not given to the machine on the input tape. Instead only the size of the input is given on the input tape. It accesses the input through an oracle. Hence, it has a special *access tape*, on which the machine can write an index, $i$, $1 \geq i \geq n$ in binary, then it is told the value of the $i$th input bit. One of Ruzzo's results was that

$$NC = \bigcup_{k=0}^{\infty} ATISP(\log^k n, \log n).$$

A resource associated with alternating Turing machines is the number of alternations that it performs. An *alternation* is a step where the machine goes from a universal state to an existential state or vice versa. Chandra, Kozen and Stockmeyer [20] looked at machines that are allowed a bounded number of alternations.

A language $L$ belongs to the class $\Sigma_i^{t,s}$ (respectively $\Pi_i^{t,s}$) for $i \geq 1$ if it is recognized by an ATM $M$ such that for all inputs, $M$ works in time $t$ and space $s$, starts in an existential (respectively universal) state and only makes $i - 1$ alternations on all computation paths.

Define the polynomial time $\Sigma_i$ languages as

$$\Sigma_i^P := \bigcup_{k=0}^{\infty} \Sigma_i^{n^k, n^k}.$$

Observe that $NP = \Sigma_1^P$ and $coNP = \Pi_1^P$. We let $\Sigma_0^P = P$.

The polynomial time hierarchy $PH$ was defined by Stockmeyer [69] as

$$PH := \bigcup_{i=0}^{\infty} \Sigma_i^P.$$

A question related to $P \overset{?}{=} NP$ is whether $PH$ collapses to a finite level, *i.e.*, if $\exists k : PH = \bigcup_{i=0}^{k} \Sigma_i^P$. A simple argument shows that if $P = NP$ then $PH = \Sigma_0$. On the other hand $PH$ could collapse and $P \neq NP$. Thus the $PH$ question is weaker than $P \overset{?}{=} NP$. Just as it is believed that $P \neq NP$ it is believed that $PH$ does *not* collapse.

## 2.3.5   Interactive Proof System Complexity Classes

Interactive proofs have been studied for a much shorter time than have the previously discussed complexity classes. The class most studied has been the equivalent of $P$ for deterministic complexity. In defining the complexity of interactive proofs we are only interested in the resources used by the verifier. Space is defined as the number of cells scanned by the verifier on its work tape. Note that cells scanned on its random tape and on the communication tapes are for free. Time is just the number of steps used by the verifier.

The prover can use whatever resources it wants. So we define $IPTISP(t, s)$ to be the class of languages that are recognizable by interactive proofs where the verifier works in time $O(t)$ and space $O(s)$. The class $IP$ is the class of languages recognized by interactive proof systems, where the verifier works in polynomial time.

We denote $pIP$ as the class of languages that are recognizable by a public-coin interactive proof system with a polynomial time verifier. Analogously, we define $pIPTISP(t, s)$. Goldwasser and Sipser [41] showed that $IP = pIP$. In fact, they showed that for some constant $k$

$$IPTIME(t) \subset pIPTIME(t^k),$$

for any $t$ that is time constructable. Hence, public coin are no restriction for polynomial time verifiers. But for other resource bounds there is a difference as shown by Condon [21]. She showed that

$$pIPTISP(poly, \log n) \subset P,$$

and she also proved that for any $t$ that is fully time constructable

$$pIPTISP(t, t) \subset IPTISP(t, \log t),$$

hence

$$IP = pIP = IPTISP(poly, \log n).$$

The first result was proved independently by Fortnow and Sipser [34]. They also proved that $LOGCFL \subseteq pIPTISP(poly, \log n)$ where $LOGCFL \subseteq NC^2$ is the class of languages log-space reducible to context-free languages [71, 72].

Condon and Lipton [22] showed that if the verifier has constant work space but infinite time then interactive proof systems can recognize any recursively enumerable language. This is true, assuming we loosen our assumption that the verifier always halts. There are some contents of the random tape for which the verifier runs indefinitely. This result is sensitive to the precise definition of which language an interactive proof system recognizes. Let us change the definition to force the verifier to reject false statements with probability at least $\frac{2}{3}$, instead of just requiring the verifier to accept with probability at most $\frac{1}{3}$. Note that, here, since we allow the verifier to run indefinitely, the two definitions are not equivalent. Indeed, Condon and Lipton [22] showed that, with this new variant of language recognition, interactive proof systems with constant work space and infinite time verifiers only recognize languages contained in $ATIME(2^{2^{O(n)}})$.

Another resource that is measured in relation to interactive proofs is the number of rounds used in the proof systems. We denote the class of languages recognized by public-coin interactive proof systems in $k$ rounds by $MA[k]$ when the prover starts and by $AM[k]$ when the verifier starts. Babai [5] showed that $MA[O(1)]$ is equal $AM[1]$, *i.e.*, to the class recognized by public-coin interactive proof systems that consist of only one message from the verifier, then one message from the prover and then some local deterministic computation by the verifier. We denote $MA = MA[1]$ and $AM = AM[1]$. Babai and Moran [10] proved that $MA \subset \Sigma_2^P \cap \Pi_2^P$ and that $AM \subset \Pi_2^P$.

Similarly, we define *MIP* to be the class of languages recognized by multiple-prover interactive proof systems.

For some corollaries of our results we need to know how many resources the prover or provers uses. We say that the honest prover *lives* in $\mathcal{C}$ if it is computing functions in $f\mathcal{C}$.

### 2.3.6 Non-Uniform Complexity Classes

A language $L$ defines a family $\{f_n \mid n \in \{0, 1, 2, \ldots\}\}$ of Boolean functions, where $f_n : \{0, 1\}^n \to \{0, 1\}$ is the characteristic function for $L \cap \{0, 1\}^n$. In other words

$$f_n(x_1, x_2, \ldots, x_n) = \begin{cases} 1 & \text{if } x_1 x_2 \cdots x_n \in L \\ 0 & \text{otherwise.} \end{cases}$$

A family of circuits $C = \{C_n \mid n \in \{0, 1, 2, \ldots\}\}$ computes $L$ if $C_n$ computes $f_n$ for all $n$. In general, we do not require that there is a relationship between the circuits in a family $C$, we say that the family is a *non-uniform* circuit family. This makes it possibly for circuits to compute languages that no TM can compute. We can required that the circuits in a family $C$ have some common relationship by requiring that a description of the circuits can be computed by a TM $M$. We say that the circuit family is *log-space uniform* if $M$ uses logarithmic space.

We define non-uniform complexity classes by restricting the size and/or depth of the circuits. We define $P/\text{poly}$ to be the class of languages that have polynomial size circuits. If we restrict our circuits to uniform polynomial size circuits we just get the class $P$. Bennet and Gill [15] showed that $BPP \subset P/\text{poly}$.

Cook [24] defined for a constant $k$, $NC^k$ to be the class of languages that have polynomial size and $O(\log^k n)$ depth uniform circuits. Hence, if $L \in NC^k$ then there exists a family of circuits $C = \{C_n \mid n = 1, 2, \ldots\}$ such that $C$ computes $L$ and there exists a constant $k'$ such that $size(C_n) = O(n^{k'})$ and $depth(C_n) = O(\log^k n)$. Furthermore, there exists a log-space Turing machine that, on input $1^n$ computes a description of $C_n$. Furthermore, he defined $NC$ as $\bigcup_{k \geq 1} NC^k$.

## 2.3.7  Complexity Classes Relative to Oracles

We define the complexity of oracle machines just like the complexity of similar machines without oracles. For example given an oracle $A \subset \Sigma^*$ the class $P^A$ consists of languages that are recognizable by polynomial time deterministic OTM with oracle $A$.

Baker, Gill and Solovay [11] introduced these concepts as a tool to show that some techniques will not settle the $P$ versus $NP$ question. They found oracles $A, B$ such that, for example, $P^A = NP^A$ and $P^B \neq NP^B$. Since many of the results in complexity theory *relativize*, meaning that they hold for an arbitrary oracle too, we know that using the techniques from these results will not settle the $P$ versus $NP$ question since we know oracles that answer the question both ways.

Here is a sample of other oracle results from [11]

$$\exists C : NP^C \neq coNP^C$$

$$\exists D : P^D = NP^D = coNP^D$$

$$\exists E : P^E \neq NP^E \text{ and } P^E = NP^E \cap coNP^E$$

$$\exists F : NP^F \neq coNP^F \text{ and } P^E \neq NP^F \cap coNP^F.$$

In this book we will describe a technique that does *not* relativize. Furthermore, we settle a question in a way that goes against previously published oracle results.

**Fact 2.3.4 (Fortnow and Sipser [35])** *There exists an oracle A such that*

$$coNP^A \nsubseteq IP^A.$$

**Fact 2.3.5 (Fortnow, Rompel and Sipser [33])** *There exists an oracle A such that*

$$coNP^A \nsubseteq MIP^A.$$

## 2.4   Counting Classes

NTMs can be looked at as defining a function instead of defining a language. Instead of being interested in determining if an accepting path exists we are interested in the number of accepting paths.

Given an NTM $M$ define $\#M : \Sigma^* \rightarrow \mathbf{N}$ by letting $\#M(x)$ equals the number of accepting paths of $M$ on input $x$. The class $\#P$ of functions introduced by Valiant in [76] is the class describable as $\#M$ for some polynomial time NTM $M$.

$$\#P := \{\#M \mid M \text{ is a polynomial time NTM}\}.$$

Recently, Toda [74] showed that there are $\#P$ problems, which are hard for the polynomial time hierarchy, in the sense that given an oracle for the problem we can recognize any language in *PH* in polynomial time.

## 2.5   Completeness

One of the most important tools to study complexity classes is the *reduction*, which was introduced into complexity theory by Cook [23], Levin [53] and Karp [50]. Reductions enable us to talk about the relative hardness of problems. So if we can reduce a problem $\mathcal{A}$ to a problem $\mathcal{B}$ with a reduction $r$ then we can say that problem $\mathcal{A}$ in some way is not harder than $\mathcal{B}$.

There are many types of reductions and we use a couple of them.

### Turing Reductions

The most general reduction is the *Turing reduction*. A problem $\mathcal{A}$ is Turing reducible ($\leq_T^P$) to a problem $\mathcal{B}$ if there exists a polynomial time deterministic OTM $M$ such that $M^{\mathcal{B}}$ solve $\mathcal{A}$. Hence, a language $L$ is Turing reducible to a language $L'$ if and only if $L \in P^{L'}$.

We say that two languages $L$ and $L'$ are *Turing equivalent* if and only if $L \leq_T^P L'$ and $L' \leq_T^P L$.

### Many-One Reductions

Another well known reduction is *many-one reduction* ($\leq_m^P$). It is a special type of Turing reduction where the OTM can ask exactly one question to the oracle and has to accept if and only if the oracle answers "yes."

### Completeness

The importance of reductions stems from the notion of *completeness* which was introduced by Cook and Levin. We say that a language $L$ is complete for a class $\mathcal{C}$ with respect to $r$-reductions if and only if $L \in \mathcal{C}$ and for all $L' \in \mathcal{C}$ we have that $L' \leq_r L$. If a problem is complete for a class it is, in a sense, one of the hardest problems in that class.

Since completeness was introduced, complete problems have been found for many complexity classes. See Garey and Johnson's excellent book [37] on completeness for an extensive list of complete problem and a more detailed discussion of completeness.

Many complete problems arise in logic. Cook [23] and Levin [53] found the first complete problem for *NP*. They showed that SAT is *NP*-complete. SAT is the set of CNF formulas $\varphi$, that have a satisfying assignment.

$$\text{SAT} := \{\varphi \mid \varphi \text{ is satisfiable}\}.$$

A variant of SAT is 3-SAT that is the same problem but where $\varphi$ is a 3-CNF formula, *i.e.*, each clause contains at most 3 literals. An easy argument [37, pages 48-49] shows that $\text{SAT} \leq_m^P 3\text{-SAT}$. Hence 3-SAT is also *NP*-complete. Since 3-SAT are simpler but include the computational difficulties of SAT we look at 3-CNF formulas instead of general CNF formulas.

When looking at 3-SAT, it can be seen [37, page 169] that if we count the number of satisfying assignments then we have a $\#P$-complete problem. Given a 3-CNF formula $\varphi$ we let $\#\varphi$ be the number of satisfying assignments. The function $\#3\text{-SAT}$ defined by $\#3\text{-SAT}: \varphi \mapsto \#\varphi$. $\#3\text{-SAT}$ is $\#P$-complete, where $\#P$-completeness is defined relative to Turing reductions.

A *PSPACE*-complete problem is TQBF (True Quantified Boolean Formula) [70, 69]. TQBF is the problem given a 3-CNF formula $\varphi$ with Boolean variables $x_1, x_2, \ldots, x_n$ to decide the truth value of

$$\forall x_1 : \exists x_2 : \cdots \odot x_n : \varphi(x_1, x_2, \ldots, x_n),$$

where $\odot = \forall$ (respectively $\odot = \exists$) if $n$ is odd (respectively even).

$$\text{TQBF} := \{\varphi \mid \forall x_1 : \exists x_2 : \cdots \odot x_n : \varphi(x_1, x_2, \ldots, x_n)\}.$$

## 2.6   Summary of Our Results

In Chapter 3 we show that every language in $P^{\#P}$ has an interactive proof. This result is from the joint work with Fortnow, Karloff and Nisan [55]. We furthermore for sake of completeness present an extension of our result due to Shamir [66], which shows that every language in *PSPACE* has an interactive proof.

In Chapter 4 we show that every language in *NEXP* has a two-prover interactive proof. This result is from the joint work with Babai and Fortnow [9].

In Chapter 5 we obtain a polynomial relationship between time-space bounded ATMs and single-prover public-coin interactive proof system with time-space bounded verifiers. This research is joint work with Fortnow [32].

Chapter 6 describes some consequences of our work in different areas of computer science.

# 3

# The Power of Interaction with One Prover

In this chapter we are going to describe single-prover interactive proof systems for the two complexity classes $\#P$ and $PSPACE$. The proof of the main Theorems 3.1.8 and 3.2.11 in this chapter follows the same outline: Given a complexity class, we arithmetize first a complete problem from the class and then give an interactive proof that solves the arithmetized problem.

## 3.1 Proof Systems for $\#P$

In this section we give interactive proofs for every language in $P^{\#P}$. This implies, using a recent result by Toda [74], that every language in $PH$ and, in particular, every language in $coNP$ has interactive proof systems. Before this result, there were only a few interactive proof systems for languages not known to belong to $NP$. Furthermore, Fortnow and Sipser had constructed an oracle, relative to which there existed a language in $coNP$ which did not have an interactive proof.

### 3.1.1 Arithmetizing of $\#P$

In this section we construct a reduction from a $\#P$-complete problem to the problem of evaluating a specific type of arithmetic expression.

The #$P$-complete problem we consider is #3SAT. Recall that #3SAT problem is: given a 3-CNF formula $\varphi$, find the number #$\varphi$ of satisfying assignments of the variables in $\varphi$. This is a well known #$P$-complete problem [37]. First, we prove a lemma that arithmetizes the problem of checking that an assignment satisfies a 3-CNF formula.

**Lemma 3.1.1** *Let $\varphi$ be a 3-CNF formula with $n$ variables $x_1, x_2, \ldots, x_n$ and $m$ clauses. Then there exists a polynomial time computable arithmetic formula $F_\varphi(x_1, x_2, \ldots, x_n)$, such that for all $\alpha_1, \alpha_2, \ldots, \alpha_n \in \{0, 1\}$:*

$$F_\varphi(\alpha_1, \alpha_2, \ldots, \alpha_n) = \begin{cases} 1 & \text{if } \varphi(\alpha_1, \alpha_2, \ldots, \alpha_n) \text{ is true} \\ 0 & \text{otherwise.} \end{cases}$$

*The size of $F_\varphi$ is at most $20m$ and $\deg(F_\varphi) = 3m$.*

**Proof:**  Let

$$\varphi = \bigwedge_{i=1}^{m} y_i^1 \vee y_i^2 \vee y_i^3,$$

where $y_i^j \in \{x_1, x_2, \ldots, x_n, \overline{x}_1, \ldots, \overline{x}_n\}$ for $i \in \{1, 2, \ldots, m\}$ and $j \in \{1, 2, 3\}$.

We observe that $\wedge$ can be expressed as multiplication in the algebraic setting. Similarly negation $\overline{x}$ and disjunction $x \vee y$ can be expressed as $1 - x$ and $1 - (1 - x)(1 - y)$ respectively. This leads to the following definition.

$$F_\varphi := \prod_{i=1}^{m} \left(1 - p_i^1 p_i^2 p_i^3\right),$$

where

$$p_i^j = \begin{cases} (1 - x_l) & \text{if } y_i^j = x_l \\ x_l & \text{if } y_i^j = \overline{x}_l. \end{cases}$$

Observe that for $\alpha_1, \alpha_2, \ldots, \alpha_n \in \{0, 1\}$:

$$(1 - p_i^1 p_i^2 p_i^3)(\alpha_1, \alpha_2, \ldots, \alpha_n) =$$
$$\begin{cases} 1 & \text{if } (\alpha_1, \alpha_2, \ldots, \alpha_n) \text{ satisfies the } i\text{th clause} \\ 0 & \text{otherwise.} \end{cases}$$

So

$$F_\varphi(\alpha_1, \alpha_2, \ldots, \alpha_n) = \begin{cases} 1 & \text{if } (\alpha_1, \alpha_2, \ldots, \alpha_n) \text{ satisfies } \varphi. \\ 0 & \text{otherwise.} \end{cases} \blacksquare$$

**Example 3.1.2** *If $\varphi$ is*

$$(x_1 \vee x_2 \vee \neg x_3) \wedge (x_4 \vee \neg x_1 \vee \neg x_2)$$

*we get that the arithmetization of $\varphi$ is*

$$(1 - (1 - x_1)(1 - x_2)x_3)(1 - (1 - x_4)x_1x_2).$$

Having $F_\varphi$, it is easy to see how to construct an arithmetic expression that counts the number of satisfying assignments. Just add up $F_\varphi(x_1, x_2, \ldots, x_n)$ for all 0 and 1 assignments to the variables.

**Definition 3.1.3** *Given $\varphi$ be a 3-CNF formula with $n$ variables $x_1, \ldots, x_n$ and $m$ clauses define the arithmetic expression $A_\varphi$ as*

$$\sum_{x_1=0}^{1} \sum_{x_2=0}^{1} \cdots \sum_{x_n=0}^{1} F_\varphi(x_1, x_2, \ldots, x_n).$$

$A_\varphi$ is an arithmetic expression with no free variables. The value of $A_\varphi$ depends on the field $\mathbf{F}$ in which it is evaluated. The value is equal to $\#\varphi$ modulo the characteristic ($\mathrm{char}(\mathbf{F})$) of $\mathbf{F}$. For example, if the field has characteristic 2 (*i.e.*, $1+1=0$) then the value of $A_\varphi$ is equal to $\#\varphi \pmod 2$.

If $\mathrm{char}(\mathbf{F}) > 2^n$ or $\mathrm{char}(\mathbf{F}) = 0$, the value equals $\#\varphi$, since there are at most $2^n$ satisfying assignments. In this chapter, we work in the finite field with $p$ elements $\mathbf{F}_p$, where $p$ is a prime in the interval $[\max(2^n, 9nm), 2\max(2^n, 9nm)]$ (such a prime always exists because of "Bertrand's Postulate" [42, Theorem 418]). We have a reduction from #3SAT to evaluating a special type of arithmetic expression.

**Lemma 3.1.4** *If $\mathrm{char}(\mathbf{F}) > 2^n$ or $\mathrm{char}(\mathbf{F}) = 0$ then the value of $A_\varphi$ in the field $\mathbf{F}$ is equal to the number of satisfying assignments of $\varphi$.*

**Proof:** Follows from the above discussion. ∎

**Remark 3.1.5** The arithmetization in this section is from $\#P$ instances to arithmetic expressions. Babai and Fortnow, in [7], arithmetize $\#P$-functions as uniform families of arithmetic expressions of a special type.

## 3.1.2    Proof Systems for $\#P$

In this section, we describe interactive proof systems for every language in $P^{\#P}$. We use the arithmetized problem that was introduced in the previous section.

We start with a protocol for the language

$$L := \{(\varphi, s) \mid \varphi \text{ is a 3-CNF formula, } s \in \mathbf{Z} \text{ and } s = \#\varphi\}.$$

Hence the input is a 3-CNF formula $\varphi$ and a number $s$ and it is the verifier's job to verify that the number of satisfying assignments of $\varphi$ is $s$. From Lemma 3.1.4, therefore this is the same as verifying that $A_\varphi = s$.

The basic idea of the protocol is to eliminate the $\Sigma$s one by one in $A_\varphi$. We denote the arithmetic expression obtained from $A_\varphi$ by eliminating the outermost $\Sigma$ by $A_1(x_1)$. Observe that in $A_1(x_1)$, $x_1$ is a free variable. Let $A_i(x_1, x_2, \ldots, x_i)$ denote the arithmetic expression obtained from $A_\varphi$ by eliminating the $i$ outermost $\Sigma$s. In other words

$$A_i(x_1, x_2, \ldots, x_i) := \sum_{x_{i+1}=0}^{1} \sum_{x_{i+2}=0}^{1} \cdots \sum_{x_n=0}^{1} F_\varphi(x_1, x_2, \ldots, x_n).$$

The protocol consists of $n$ rounds. During the protocol the verifier is in the following situation at the beginning of the $i$th round:

> It has to verify that $A_{i-1}(\alpha_1, \alpha_2, \ldots, \alpha_{i-1})$ is equal to some value $\beta_{i-1} \in \mathbf{F}_p$ for $\alpha_1, \alpha_2, \ldots, \alpha_{i-1}$ in $\mathbf{F}_p$.

Observe that at the start of the protocol the verifier has to verify that $A_0$ (equivalent to $A_\varphi$) is equal to $s$. Hence if we let $\beta_0$ equal $s$ then the verifier is in exactly the above situation for $i = 1$.

The verifier's goal in the $i$th round is to choose $\alpha_i$ and $\beta_i$ such that $A_i(\alpha_1, \alpha_2, \ldots, \alpha_i) = \beta_i$ if and only if $A_{i-1}(\alpha_1, \alpha_2, \ldots, \alpha_{i-1}) = \beta_{i-1}$. We describe how the verifier chooses $\alpha_i$ and $\beta_i$ shortly. First, we observe that if the verifier achieves its goal in each round then after $n$ rounds it only has to verify that $A_n(\alpha_1, \alpha_2, \ldots, \alpha_n) = \beta_n$ to verify that $A_\varphi = s$. From the definition of $A_n(\alpha_1, \alpha_2, \ldots, \alpha_n)$, this is the same as verifying that $F_\varphi(\alpha_1, \alpha_2, \ldots, \alpha_n) = \beta_n$, where $F_\varphi$ is the formula from Lemma 3.1.1. The verifier can, therefore, do this verification herself.

It remains for the verifier to choose proper $\alpha_i$ and $\beta_i$. The following observations provide the key to this.

1. $q(x_i) := A_i(\alpha_1, \alpha_2, \ldots, \alpha_{i-1}, x_i)$ is a polynomial of degree at most $3m$ in $x_i$.

2. $A_{i-1}(\alpha_1, \alpha_2, \ldots, \alpha_{i-1}) = \sum_{x_i=0}^{1} A_i(\alpha_1, \alpha_2, \ldots, \alpha_{i-1}, x_i)$.

The idea is to let the verifier ask for $q$. The prover gives the verifier $q'$ which may or may not be $q$. The verifier checks that the polynomial $q'$ it is given satisfies the second observation, *i.e.*, that $q'(0) + q'(1) = \beta_{i-1}$. It then chooses $\alpha_i$ uniformly at random in $\mathbf{F}_p$ and lets $\beta_i = q'(\alpha_i)$. Observe that if the prover gave the verifier the polynomial $q$ then in the next round $\beta_i = q(\alpha_i) = A_i(\alpha_1, \alpha_2, \ldots, \alpha_i)$. A cheating prover has only a slim chance of succeeding, since it has to give the verifier $q'(x_i)$ which does not equal to $q$, since $q'(0) + q'(1) = \beta_{i-1}$ but $q(0) + q(1) \neq \beta_{i-1}$. Suppose that the prover cheats and gives a $q' \neq q$. In this case, $q$ and $q'$ agree on at most $3m$ points in $\mathbf{F}_p$, since both are polynomials of degree at most $3m$. For a randomly and uniformly picked $\alpha_i$, there is an at most $\frac{3m}{p}$ chance that $q'(\alpha_i) = q(\alpha_i)$. In other words there is, at most, a $\frac{3m}{p}$ chance that a cheating prover can make the verifier miss its goal. See Figure 3.1 for all details of the protocol.

We show next that Protocol 1 is an interactive proof system for $L$. First we prove that if at the beginning of the $i$th round the verifier has to verify a false statement then there are only a few bad choices of $\alpha_i$ such that at the end of the round the verifier has to verify a true statement.

**Lemma 3.1.6** *For all $i$ and all $\alpha_1, \alpha_2, \ldots, \alpha_{i-1}, \beta \in \mathbf{F}_p$ such that $A_{i-1}(\alpha_1, \alpha_2, \ldots, \alpha_{i-1}) \neq \beta$ and for all polynomials $q'$ over $\mathbf{F}_p$ of degree at most $3m$, $q'(0) + q'(1) = \beta$ implies that*

$$|\{\alpha \in \mathbf{F}_p \mid q'(\alpha) = A_i(\alpha_1, \alpha_2, \ldots, \alpha_{i-1}, \alpha)\}| \leq 3m.$$

---

**Protocol 1**

**P→V:** a prime $p \in [\max(2^n, 9nm), 2\max(2^n, 9nm)]$ and a proof that $p$ is a prime. From Pratt [62] we know that there exists a polynomial length proof showing that $p$ is indeed a prime.

**V:** $\beta_0 \leftarrow s$.

   **for** $i = 1$ **to** $n$ **do**

   **P→V:** $q_i$ a polynomial over $\mathbf{F}_p$ of degree at most $3m$.

   **V:** Checks that $q_i(0) + q_i(1) = \beta_{i-1}$ **else Halt** and **Reject**.

   **V:** Chooses $\alpha_i$ uniformly and randomly in $\mathbf{F}_p$.

   **V:** $\beta_i \leftarrow q_i(\alpha_i)$

   **end**

**V: if** $\beta_n = F_\varphi(\alpha_1, \alpha_2, \ldots, \alpha_n)$ **then**

   **Halt** and **Accept**

**else**

   **Halt** and **Reject**

---

Figure 3.1: A protocol for L.

**Proof:** Let $q(x)$ be the polynomial of degree at most $3m$ obtained by setting $q(x) := A_i(\alpha_1, \alpha_2, \ldots, x)$. Since $q(0) + q(1) = A_{i-1}(\alpha_1, \alpha_2, \ldots, \alpha_{i-1}) \neq \beta$ we have $q'(0) + q'(1) \neq q(0) + q(1)$ and therefore $q' \neq q$. Elements in $\mathbf{F}_p$ where $q'$ and $q$ coincide are roots in the non-zero polynomial $q' - q$ which has degree at most $3m$. But any non-zero polynomial of degree at most $3m$ has at most $3m$ roots. This completes the proof. ∎

The above lemma lies at the heart of most of the results in this book.

**Lemma 3.1.7** *Protocol 1 satisfies the following statements:*

1. *If $(\varphi, s) \in L$ then there exists a prover such that the verifier accepts always.*

2. *If $(\varphi, s) \notin L$ then for all provers, the verifier accepts with probability at most $\frac{3nm}{p}$.*

*3. The verifier runs in time $O(mn \log^2 p) = O(mn^3)$.*

*and therefore $L \in IP$.*

**Proof:**

1. The prover gives $q_i(x) := A_i(\alpha_1, \alpha_2, \ldots, \alpha_{i-1}, x)$ in the $i$th round. The two observations above imply that the verifier always accepts.

2. Assume that $A_0 \neq \beta_0$. Fix a prover that tries to convince the verifier that $A_0 = \beta_0$. We assume without loss of generality that the prover never induces the verifier to reject until the last step in the protocol (this is because the probability of rejection by the verifier will only decrease). We say that the prover *succeeds in round $i$* if and only if $A_i(\alpha_1, \alpha_2, \ldots, \alpha_i) = \beta_i$ but $A_{i-1}(\alpha_1, \alpha_2, \ldots, \alpha_{i-1}) \neq \beta_{i-1}$. Observe that for the verifier to accept, the prover has to succeed in at least one round. Hence

$$\Pr[\text{The verifier accepts}] \leq \sum_{i=1}^{n} \Pr[\text{The prover succeeds in round } i].$$

   For the prover to succeed in the $i$th round, the verifier must have chosen $\alpha_i$ such that $q_i(\alpha_i) = A_i(\alpha_1, \alpha_2, \ldots, \alpha_i)$. From Lemma 3.1.6 there are at most $3m$ elements in $\mathbf{F}_p$ for which this happens. Since the verifier chooses $\alpha_i$ uniformly and randomly from $\mathbf{F}_p$, we know that the probability of the prover succeeding in round $i$ is no more than $\frac{3m}{p}$. Hence

$$\Pr[\text{The verifier accepts}] \leq \sum_{i=1}^{n} \frac{3m}{p} = \frac{3nm}{p}.$$

3. The verifier is only doing $n$ evaluations of polynomials of degree at most $3m$ and an evaluation of $F_\varphi$. This can be done by $O(nm)$ arithmetic operations, each of which can be done by $O(\log^2 p)$ bit operations. ∎

The main result of this section now follows.

**Theorem 3.1.8**

$$P^{\#P} \subset IP.$$

**Proof:** We observe that $P^{\#P} = P^{\#3SAT}$ since #3SAT is #P-complete. Let $M$ be a polynomial time oracle Turing machine with oracle #3SAT. We construct an interactive proof system that recognizes exactly the same language as $M$. The verifier simulates $M$ on input $x$. Whenever $M$ asks an oracle question $\#\varphi$ the verifier asks the same question to the prover. After getting the answer, $s$, the verifier asks the prover to prove to it using Protocol 1 that $(\varphi, s) \in L$. It rejects $x$ if it rejects $(\varphi, s)$ in the subprotocol.

If $x \in L(M)$ then a prover who answers every question truthfully makes the verifier accept. If $x \notin L(M)$ then the prover has to cheat on at least one question and therefore because of Lemma 3.1.7 the probability that the verifier accepts is less that $\frac{1}{3}$. Hence the verifier recognizes $L(M)$.

Since $M$ works in polynomial time and since the verifier in Protocol 1 works in polynomial time then the verifier works in polynomial time.  ▌

Goldwasser and Sipser [41] have shown that one can convert any interactive proof to one where the verifier uses public-coins, *i.e.*, the verifier can only flip coins "in front of" the prover. Goldreich, Mansour and Sipser [38] have shown how to modify an interactive proof system so that for true instances the verifier is convinced with probability one. Both of these properties already hold for our protocol.

For some of the applications of this result it is important to know what the complexity of the honest prover is. The next theorem states that the prover only has to be in $\#P$.

**Theorem 3.1.9** $P^{\#P}$ *has interactive proof systems where the prover lives in* $\#P$.

**Proof:** We have to change our protocol a little such that the only thing the prover has to do is to evaluate arithmetic expressions, $A_i(x_1, x_2, \ldots, x_i)$ at points $(\alpha_1, \alpha_2, \ldots, \alpha_i)$ in $\mathbf{F}_p^i$. Lemma 3.1.10, below, states that a prover in $\#P$ can do this.

The idea is to let the verifier not ask for a polynomial $q$ but instead ask for $q$ to be evaluated at $3m + 1$ points. Since the degree of $q$ is at most $3m$ the

---

**Protocol 2**
  **V:** $\beta_0 \leftarrow s$.

    **for** $i = 1$ **to** $n$ **do**
        **for** $j = 0$ **to** $3m$ **do**
            **P→V:** A $\gamma_j \in \mathbf{F}_p$.
        **end**
        **V:** Interpolate a polynomial $q_i$ over $\mathbf{F}_p$ of degree at most $3m$ such that
            $q_i(j) = \gamma_j$ for $j \in \{0, 1, \ldots, m\}$.
        **V:** Check that $q_i(0) + q_i(1) = \beta_{i-1}$ **else Halt** and **Reject**.
        **V:** Choose $\alpha_i$ uniformly and randomly in $\mathbf{F}_p$.
        **V:** $\beta_i \leftarrow q_i(\alpha_i)$
    **end**
  **V:** **if** $\beta_n = F_\varphi(\alpha_1, \alpha_2, \ldots, \alpha_n)$ **then**
    **Halt** and **Accept**
  **else**
    **Halt** and **Reject**

---

Figure 3.2: A protocol for L with prover in $\#P$.

verifier can then interpolate $q$. For all the details of the protocol see Figure 3.2. ∎

**Lemma 3.1.10** *Given $(\alpha_1, \alpha_2, \ldots, \alpha_i)$ in $\mathbf{F}_p^i$, there exists a polynomial time nondeterministic Turing machine $M$ such that for $x = (\varphi, i, \alpha_1, \alpha_2, \ldots, \alpha_i)$ we have that*

$$A_i(\alpha_1, \alpha_2, \ldots, \alpha_i) = \#M(x) \pmod{p}.$$

**Proof:** $M$ guesses first nondeterministically $n - i$ bits $x_{i+1}, x_{i+2}, \ldots, x_n$. Then it evaluates $F_\varphi$ at $(\alpha_1, \alpha_2, \ldots, \alpha_i, x_{i+1}, x_{i+2}, \ldots, x_n)$ and if this number $\beta$ is not 0, then $M$ makes $\beta$ accepting paths, otherwise it makes 1 rejecting path. ∎

# 3.2  Proof Systems for *PSPACE*

In this section we follow the same pattern as in Section 3.1, but for the complexity class *PSPACE* instead of #*P*. We construct an arithmetization of a *PSPACE*-complete problem. This arithmetization is then used to construct interactive proof systems for *PSPACE*.

Our *PSPACE*-complete problem is TQBF (see section 2.5).

## 3.2.1  Arithmetizing of *PSPACE*

A straightforward arithmetization does not allow the techniques for constructing interactive proof systems for #*P* to be used. So we have to use a trick in the arithmetization that allows us to use the techniques anyway. We start with the straightforward arithmetization of QBFs and discuss the problem with it.

First notice that the operator $\prod_{i=0}^{1} \alpha_i$ corresponds to an AND of $\alpha_0$ and $\alpha_1$ if $\alpha_0, \alpha_1 \in \{0,1\}$. So $\prod$ operators correspond to $\forall$ quantifiers. We need an arithmetic operator that corresponds to $\exists$ quantifier as well. So we define

$$\coprod_{i=0}^{1} \alpha_i := 1 - \prod_{i=0}^{1} (1 - \alpha_i).$$

It is easy to see that $\coprod_{i=0}^{1} \alpha_i$ corresponds to an OR of $\alpha_0$ and $\alpha_1$. This gives the following arithmetization of QBFs.

**Definition 3.2.1** *Let $\varphi$ be a 3-CNF formula with $n$ variables $x_1, x_2, \ldots, x_n$ and $m$ clauses. Define*

$$B_\varphi = \prod_{x_1=0}^{1} \coprod_{x_2=0}^{1} \prod_{x_3=0}^{1} \cdots \bigodot_{x_n=0}^{1} F_\varphi(x_1, x_2, \ldots, x_n),$$

*where $\bigodot = \prod$ (respectively $= \coprod$) if $n$ is odd (respectively even). And $F_\varphi$ is from Lemma 3.1.1.*

**Lemma 3.2.2** *Let $\varphi$ be a 3-CNF formula then*

$$B_\varphi = \begin{cases} 1 & \text{if } \varphi \in TQBF \\ 0 & \text{otherwise} \end{cases}$$

**Proof:** Follows from the definition of $\prod, \coprod$ and Lemma 3.1.1. ∎

The problem with this arithmetization is that we can not use our interactive proof technique. Define $B_i(x_1, x_2, \ldots, x_i)$ as $B_\varphi$ without the $i$ outermost operators (the dependence on $\varphi$ is implicit). For example

$$B_3(x_1, x_2, x_3) := \coprod_{x_4=0}^{1} \prod_{x_5=0}^{1} \coprod_{x_6=0}^{1} \cdots \bigodot_{x_n=0}^{1} F_\varphi(x_1, x_2, \ldots, x_n).$$

For our technique from the $\#P$ case to work, we look at $B_i(\alpha_1, \alpha_2, \ldots, \alpha_{i-1}, x_i)$ for $\alpha_1, \alpha_2, \ldots, \alpha_{i-1} \in \mathbf{F}_p$. This is, as in the $\#P$-case, a polynomial but since $\deg_{x_1} B_i(x_1, x_2, \ldots, x_i)$ is twice the $\deg_{x_1} B_{i+1}(x_1, x_2, \ldots, x_{i+1})$ we get that $\deg_{x_1} B_1(x_1)$ can be as high as $2^{n-1}3m$. So there is no way the verifier can even read this polynomial. So we need a trick to keep the degree low. Shamir in [66] introduced dummy variables to take care of this problem. Instead we use an idea by Shen [67] since Shen's trick makes the interactive proof systems simple and similar to the protocol for $\#P$.

We introduce operators $\mathcal{R}x_i$ for $i \in \{1, 2, \ldots, n\}$ on polynomials that take a polynomial $p(x_1, x_2, \ldots, x_n)$ and reduce it to a polynomial that is linear in $x_i$.

**Definition 3.2.3** *Define $\mathcal{R}x_i : \mathbf{F}[x_1, x_2, \ldots, x_n] \to \mathbf{F}[x_1, x_2, \ldots, x_n]$ as for $p \in \mathbf{F}[x_1, x_2, \ldots, x_n]$*

$$\mathcal{R}x_i(p)(x_1, x_2, \ldots, x_n) := p(x_1, x_2, \ldots, x_n) \pmod{x_i^2 - x_i}.$$

$\mathcal{R}x_i$ just makes any occurrence of $x_i^j$ for $j > 1$ into an occurrence of $x_i$.

**Example 3.2.4** $\mathcal{R}x_2(x_1^2 x_2^2 + x_2^3 + x_2) = x_1^2 x_2 + 2x_2$.

We arithmetize a QBF by adding $\mathcal{R}x_i$ such that the degrees are kept low.

**Definition 3.2.5** *Define the arithmetic expressions $C_n, C_{n-1}, \ldots, C_0$ inductively as*

$$C_n(x_1, x_2, \ldots, x_n) = F_\varphi(x_1, x_2, \ldots, x_n)$$

*and for $i < n$:*

$$C_i(x_1, x_2, \ldots, x_i) = \bigodot_{x_{i+1}=0}^{1} \mathcal{R}x_1 \, \mathcal{R}x_2 \cdots \mathcal{R}x_{i+1} \, C_{i+1}(x_1, x_2, \ldots, x_{i+1}),$$

*where $\odot = \prod$ (respectively $= \coprod$) if $i + 1$ is odd (respectively even).*

We denote $C_0$ also by $C_\varphi$.

**Example 3.2.6** *If $n = 3$ and therefore the QBF looks like*

$$\forall x_1 \exists x_2 \forall x_3 \, \varphi(x_1, x_2, x_3)$$

*then*

$$C_\varphi = \prod_{x_1=0}^{1} \mathcal{R}x_1 \coprod_{x_2=0}^{1} \mathcal{R}x_1 \, \mathcal{R}x_2 \prod_{x_3=0}^{1} \mathcal{R}x_1 \, \mathcal{R}x_2, \mathcal{R}x_3 \, F_\varphi(x_1, x_2, x_3).$$

We show that $C_\varphi$ is indeed an arithmetization of TQBF. But, first note that this arithmetization takes care of the degree problem. For $i \in \{0, 1, \ldots, n\}$ and $j \in \{0, 1, \ldots, i+2\}$ let $C_{ij}(x_1, x_2, \ldots, x_i)$ be $C_i(x_1, x_2, \ldots, x_i)$ without the $j$ outermost operators. Observe that $C_{i0} = C_i$ and $C_{i,i+2} = C_{i+1}$. The degree of $x_k$ for $k \in \{1, 2, \ldots, i\}$ is bounded by

$$\begin{cases} 2 & \text{in } C_{i0}. \\ 1 & \text{in } C_{i1}. \\ 2 & \text{in } C_{ij} \text{ for } i \in \{0, 1, \ldots, n-2\} \text{ and } j \in \{2, 3, \ldots, i+2\}. \\ 3m & \text{in } C_{n-1,j} \text{ for } j \in \{2, 3, \ldots, i+2\}. \end{cases}$$

**Lemma 3.2.7** *For all $i \in \{0, 1, \ldots, n\}$ and all $\alpha_1, \alpha_2, \ldots, \alpha_i \in \{0, 1\}$*

$$C_i(\alpha_1, \alpha_2, \ldots, \alpha_i) = B_i(\alpha_1, \alpha_2, \ldots, \alpha_i).$$

**Proof:** This is proved by downward induction on $i$.

$i = n$ Follows from the definitions that

$$C_n(\alpha_1, \alpha_2, \ldots, \alpha_n) = F_\varphi(\alpha_1, \alpha_2, \ldots, \alpha_n)$$

and

$$B_n(\alpha_1, \alpha_2, \ldots, \alpha_n) = F_\varphi(\alpha_1, \alpha_2, \ldots, \alpha_n).$$

$i < n$ Observe that for any polynomial $p(x_1, x_2, \ldots, x_n)$ we have that

$$\mathcal{R}x_i(p)(\alpha_1, \alpha_2, \ldots, \alpha_n) = p(\alpha_1, \alpha_2, \ldots, \alpha_n)$$

since $0^2 = 0$ and $1^2 = 1$. This implies that for $j \in \{1, 2, \ldots, i+1\}$

$$C_{ij}(\alpha_1, \alpha_2, \ldots, \alpha_i) = \mathcal{R}x_j\, C_{i,j+1}(\alpha_1, \alpha_2, \ldots, \alpha_i)$$

and by induction we get that

$$C_{i1}(\alpha_1, \alpha_2, \ldots, \alpha_i) = C_{i+1}(\alpha_1, \alpha_2, \ldots, \alpha_i).$$

So

$$
\begin{aligned}
C_i(\alpha_1, \alpha_2, \ldots, \alpha_i) &= \bigodot_{x_{i+1}=0}^{1} C_{i1}(\alpha_1, \alpha_2, \ldots, \alpha_i, x_{i+1}) \\
&= \bigodot_{x_{i+1}=0}^{1} C_{i+1}(\alpha_1, \alpha_2, \ldots, \alpha_i, x_{i+1}) \\
&= \bigodot_{x_{i+1}=0}^{1} B_{i+1}(\alpha_1, \alpha_2, \ldots, \alpha_i, x_{i+1}) \\
&= B_i(\alpha_1, \alpha_2, \ldots, \alpha_i)
\end{aligned}
$$

This completes the proof. ∎

This lemma gives the analog to Lemma 3.1.4 for *PSPACE*. We have a reduction from instances of TQBF to instances of a special type of arithmetic expressions.

**Corollary 3.2.8** *Let $\varphi$ be a 3-CNF formula. Then*

$$C_\varphi = \left\{ \begin{array}{ll} 1 & \text{if } \varphi \in TQBF \\ 0 & \text{otherwise.} \end{array} \right.$$

**Proof:** From Lemma 3.2.7 and Lemma 3.2.2. ∎

**Protocol 3**

$\mathbf{P} \rightarrow \mathbf{V}$: a prime $p \in [3n(3m+n), 6n(3m+n)]$ and a proof that $p$ is a prime.

   $\mathbf{V}$: $\beta \leftarrow 1$.

      **for** $i = 0$ **to** $n - 1$ **do**

      $\mathbf{P} \rightarrow \mathbf{V}$: $q$, a linear polynomial over $\mathbf{F}_p$.

         $\mathbf{V}$: If $i$ is odd (respectively even) then check that $\prod_{x_i=0}^{1} q(x_i) = \beta$

            (respectively $\coprod_{x_i=0}^{1} q(x_i) = \beta$) **else Halt** and **Reject**.

         $\mathbf{V}$: Choose $\alpha_{i+1}$ uniformly and randomly in $\mathbf{F}_p$.

         $\mathbf{V}$: $\beta \leftarrow q(\alpha_{i+1})$

            **for** $j = 1$ **to** $i + 1$ **do**

            $\mathbf{P} \rightarrow \mathbf{V}$: $q$ a polynomial over $\mathbf{F}_p$ with degree at most 2 (respectively $3m$) if $i < n$ (respectively $i = n$).

               $\mathbf{V}$: Check that $\mathcal{R}x_j(q)(\alpha_j) = \beta$ **else Halt** and **Reject**.

               $\mathbf{V}$: Choose a new $\alpha_i$ uniformly and randomly in $\mathbf{F}_p$.

               $\mathbf{V}$: $\beta \leftarrow q(\alpha_i)$

         **end**

      **end**

   $\mathbf{V}$: **if** $\beta = F_\varphi(\alpha_1, \alpha_2, \ldots, \alpha_n)$ **then**

      **Halt** and **Accept**

   **else**

      **Halt** and **Reject**

Figure 3.3: A protocol for TQBF.

## 3.2.2  Proof Systems for *PSPACE*

Given the arithmetization of TQBF, it is straightforward to apply the ideas from the protocols for $\#P$. The verifier eliminates the arithmetic operators one by one. The operators $\prod$ and $\coprod$ work similarly to the $\sum$ in the $\#P$-proofs. The only minor problem is the presence of $\mathcal{R}x_i$ operators, but the general ideas can still be used. For the complete protocol see Figure 3.3.

Assume that the verifier has to verify that

$$\mathcal{R}x_j(C_{ij})(\alpha_1, \alpha_2, \ldots, \alpha_i) \overset{?}{=} \beta.$$

Observe that $q(x_j) := C_{ij}(\alpha_1, \alpha_2, \ldots, \alpha_{j-1}, x_j, \alpha_{j+1}, \ldots, \alpha_i)$ is a polynomial of degree at most 2 ($3m$ if $i = n - 1$) and $\mathcal{R}x(q)(\alpha_j) = \beta$. All the verifier has to do is to get a polynomial $q'$ from the prover and verify that $\mathcal{R}x(q')(\alpha_i) = \beta$, choose a new $\alpha'_j$ uniformly and randomly from $\mathbf{F}_p$, let the new $\beta'$ be $q(\alpha'_j)$ and then verify that

$$C_{ij}(\alpha_1, \alpha_2, \ldots, \alpha_{j-1}, \alpha'_j, \alpha_{j+1}, \ldots, \alpha_i) \stackrel{?}{=} \beta'.$$

As in the #P-protocol if the verifier is trying to verify a false statement, *i.e.,*

$$\mathcal{R}x_j(C_{ij})(\alpha_1, \alpha_2, \ldots, \alpha_i) \neq \beta$$

then there are only a few bad choices of $\alpha'_j$ such that

$$C_{ij}(\alpha_1, \alpha_2, \ldots, \alpha_{j-1}, \alpha'_j, \alpha_{j+1}, \ldots, \alpha_i) = \beta'.$$

The argument is exactly the same as in Lemma 3.1.6. So we get

**Lemma 3.2.9** *For all polynomials $C(x)$ of degree at most $d$, and all $\alpha, \beta \in \mathbf{F}_p$ such that $\mathcal{R}x_i(C)(\alpha) \neq \beta$ and for all polynomials $q$ over $\mathbf{F}_p$ of degree at most $d$, $\mathcal{R}x_i(q)(\alpha) = \beta$ implies that*

$$|\{\alpha \in \mathbf{F}_p \mid q(\alpha) = C(\alpha)\}| \leq d.$$

Now we can prove that Protocol 3 is an interactive proof system for TQBF.

**Lemma 3.2.10** *Protocol 3 satisfies the following statements:*

1. *If $\varphi \in TQBF$ then there exists a prover such that the verifier accepts always.*

2. *If $\varphi \notin TQBF$ then for all provers, the verifier accepts with probability at most $\frac{n(3m+n)}{p}$.*

3. *The verifier works in time $O((n^2 + nm) \log^2 n)$.*

*and therefore $TQBF \in IP$.*

**Proof:**

1. This is obvious.

2. This is the same argument as in Lemma 3.1.7 and using lemma 3.2.9 and similar Lemmas for $\prod$ and $\underset{\cdot}{\prod}$. The upper bound on the error probability follows from the fact that the error in each round is bounded by $\frac{d}{p}$, where $d$ is the degree of the polynomials used in that round. The sum of the degrees is $n(3m + n)$.

3. The verifier uses $O(n^2 + nm)$ arithmetic operations.    ∎

Since TQBF is *PSPACE*-complete we get that

**Theorem 3.2.11 (Shamir [66])**

$$PSPACE \subset IP.$$

From the result by Papadimitriou [58] that $IP \subset PSPACE$ we know that $IP = PSPACE$. Furthermore, Feldman [29] showed that the honest prover lives in *PSPACE*.

# 4

---

# The Power of Interaction with Two Provers

In this chapter we show that any language in *NEXP* has a two-prover inter-active proof system. In the spirit of the previous chapter, we start out by arithmetizing *NEXP* computations and then give an interactive proof based on the arithmetization, using ideas from the single prover case. It turns out that we need to be able to test if a multi-variate function is multilinear. In Section 4.4 we construct a test for multilinearity. This test is of independent interest in the field of program testing.

## 4.1 Another Characterization of *MIP*

Fortnow, Rompel and Sipser in [33] gave another characterization of *MIP* in terms of probabilistic oracle Turing machines.

Let $M$ be a probabilistic Turing machine with access to an oracle $O$. We define the languages $L$ that can be recognized by these machines as follows:

We say that $L$ is recognized by $M$ if and only if there is an oracle $O$ such that

1. For every $x \in L$, $M^O$ accepts $x$ with probability $> \frac{2}{3}$.

2. For every $x \notin L$ and for all oracles $O'$, $M^{O'}$ accepts with probability $< \frac{1}{3}$.

We say that $L$ has a *oracle proof system*.

One way to think about this model is that the oracle convinces $M$ to accept. This differs from the single-prover interactive protocol model in that the oracle must be fixed before the queries are made while in an interactive protocol the prover may let its future answers depend on previous questions.

**Theorem 4.1.1 (Fortnow, Rompel and Sipser [33])** *L is accepted by a polynomial time probabilistic oracle machine if and only if L is accepted by a multiple-prover interactive protocol.*

**Proof:**

1. Suppose $L$ is accepted by a probabilistic oracle machine $M$ in $n^k$ steps. We give a simulation using $mn^{k+1}$ provers, where $m = \lceil 2k \log n \rceil$. The verifier $V$ simulates $M$ and whenever $M$ asks an oracle question, $V$ asks each question to $m$ different provers, selected uniformly at random from among the provers not yet queried. If any two provers give a different answer to the same question, $V$ rejects; otherwise it substitutes the common answer for the oracle answer. $V$ keeps track of oracle questions and if $M$ asks the same question twice then the verifier already knows the answer and uses that answer instead of asking another prover. There can be at most $n^k$ questions so at most $mn^k$ of the $mn^{k+1}$ provers are never queried. The verifier accepts if and only if $M$ accepts.

    (a) If $x \in L$ then the provers can convince $V$ by just responding as the oracle would.

    (b) If the provers could convince $V$ to accept $x$ we create an oracle $O$ to convince $M$ to accept by answering the oracle questions as the respective prover does. Since each prover gets only one question, it may as well decide in advance an answer to every question. Its decision determines a function $f_i$ on the set $Q$ of possible questions. We merge these functions to a single oracle $O$ using the majority answer.

To a given question, the probability that all the provers queried by $V$ belong to the minority is less than $(2-\epsilon)^{-m}$ (where $\epsilon$ accounts for the negligible fraction of provers already queried). We have chosen the value of $m$ such that even the fraction $n^k(2-\epsilon)^{-m}$ is negligible. Therefore, with high probability, the oracle agrees with the provers on all questions asked.

Note that if we only had one prover and we asked all the oracle questions to this prover, it might not be consistent in its responses. Instead it could base them on earlier questions and answers.

2. Suppose $L$ is accepted by a multiple-prover interactive protocol. Then define $M$ as follows: $M$ simulates $V$ and remembers all messages. When $V$ sends the $j$th message to the $i$th prover, $M$ asks the oracle the question $(i, j, \ell, \beta_{i1}, \ldots, \beta_{ij})$ properly encoded and uses the response as the $\ell$th bit of the $j$th response from prover $i$ where $\beta_{i1}, \ldots, \beta_{ij}$ is everything prover $i$ has seen to that point.

   (a) If $x \in L$ then the oracle $O$ could convince $M$ to accept by encoding each prover's answer to each question.

   (b) If an oracle $O$ could convince $M$ to accept a string $x$ then the provers could convince the verifier to accept by using that $O$ to create their responses. ∎

This theorem gives a natural model that is equivalent to multiple-prover interactive proof systems and can be used to prove theorems about them, for example:

**Theorem 4.1.2 (Ben-Or, Goldwasser, Kilian and Wigderson [14])** *If a language $L$ is accepted by a multiple-prover interactive protocol then $L$ is accepted by a two-prover interactive protocol.*

**Proof:** Let $M$ be a probabilistic oracle machine that runs in time $n^i$. Have the verifier of the two-prover protocol ask all the oracle questions of the first prover, then pick one of the questions asked at random and verify the answer with the second prover. The probability that cheating provers are not caught this way is at most $(1 - n^{-i})$. Repeat this process $n^{i+1}$ times to reduce error probability below $e^{-n}$. ∎

**Theorem 4.1.3 (Fortnow, Rompel and Sipser [33])**

$$MIP \subseteq NEXP.$$

**Proof:**    Given a language $L$ in $MIP$. Let $M$ be a polynomial time probabilistic oracle Turing machine that recognizes $L$. Note that since $M$ is a polynomial time machine it can only ask a polynomial number of polynomial-sized questions of the oracle and hence it depends only on an exponential sized subset of the oracle. Construct an exponential time nondeterministic Turing machine $N$ that recognizes $L$ by first guessing the exponential sized oracle $A$ and then computing the probability $p$, that $M$ accepts the input with when given oracle $A$. If $p \geq \frac{2}{3}$ then $N$ accepts else $N$ rejects. Clearly, $L(N) = L(M)$ which completes the proof.  ∎

## 4.2   Arithmetization of *NEXP* Computation

This section, as well as the next two, is devoted to proving that any language in *NEXP* has a two-prover interactive proof system. Because $MIP \subseteq NEXP$ we get that $NEXP = MIP$. As in the single-prover case, in this section, we give an arithmetization of *NEXP* computation.

We start by outlining the ideas behind our proof. Let $L$ be a language in *NEXP* and let $M$ be a nondeterministic exponential time Turing machine that recognizes $L$. Look at the tableau describing the computation of $M$ on input $x$. Convert this to a 3-CNF formula as in the proof of the Cook-Levin theorem. There are an exponential number of variables and an exponential number of clauses. However, the clauses are easily definable; in other words, there exists a polynomial time function $f_x(i)$ that describes the variables in clause $i$. Thus,

$$L(M) \;=\; \{x | \text{there is an assignment of variables that for all}$$
$$i \text{ satisfies the clause that } f_x(i) \text{ describes}\}.$$

Suppose we were given an assignment $A$ as an oracle; for a fixed $i$ we can check in polynomial time if $A$ satisfies the clause described by $f_x(i)$.

As outlined in Section 4.1, we can create predetermined, though untrustworthy functions (oracles) using multiple-prover protocols. We can use the multiple provers to create $A$. (An easy implementation of this in our context will be given in Section 4.3.1.)

The next thing to do is to ask if $A$ satisfies the clause described by $f_x(i)$ for all $i$. However, we cannot immediately do such universal quantification with multiple provers. The obvious "statistical approach," replacing the "for all $i$" with "for most $i$" clearly fails, since for an assignment $A$, there may be only one clause that is not satisfied. Furthermore, there are an exponential number of clauses, so we are not able to find the unsatisfied clause in polynomial time.

We might try handling the universal quantification with the techniques described in Chapter 3, but these results do not relativize and $A$ may not have the proper algebraic properties necessary for this proof.

We need further reduction of the problem, involving an arithmetization of the fact that $f_x(i)$ is polynomial time computable.

Essentially Lemma 4.2.1 states that the number of accepting computations of any nondeterministic polynomial time Turing machine $M$ can be expressed as a simple arithmetic expression $E$. Note that the same expression is used for all inputs of length $n$, *i.e.*, there are $n$ free variables $x_1, x_2, \ldots, x_n$ in $E$, and for each $x \in \{0,1\}^n$ the value of $E(x)$ is the number of accepting computations of $M$.

**Lemma 4.2.1** *If $M$ is a nondeterministic polynomial time Turing machine then there exists a constant $c$ such that for every $n$ there exists a polynomial time computable arithmetic formula $P_n$ in $n + n^c$ variables and of degree at most $n^c$ such that for every $x \in \{0,1\}^n$*

$$\sum_{z \in \{0,1\}^{n^c}} P_n(x, z) = \text{the number of accepting computations of } M \text{ on input } x.$$

**Proof:** Let $\varphi_n(x, z)$ be the 3-CNF formula from the proof of the Cook-Levin Theorem.

Let us review the proof of the Cook-Levin Theorem. Since $M$ is a nondeterministic polynomial time Turing machine, there exists a constant $c'$ such that $M$ works in time $O(n^{c'})$. Given $n$, look at a computation path of $M$ on an input $x$ of length $n$, as a sequence of IDs of $M$. The length of the sequence is, at most, $O(n^{c'})$. We can encode every ID by a binary code using, at most, $O(n^{c'})$ bits, since $M$ uses at most $O(n^{c'})$ tape cells. Therefore, we can encode the whole sequence using $O(n^{2c'})$ bits. These bits are the variables $z_1, z_2, \ldots, z_{n'}$ in $\varphi_n(x, z)$, where $n' = O(n^{2c'})$. On the other hand, any assignment of $z$ gives us a sequence of IDs. For some assignments the sequence corresponds to an accepting computation path, for others it corresponds to a rejecting computation path and for the rest the sequence does not correspond to a computation path at all. We can construct clauses that are satisfiable only if the assignment of the variables correspond to an accepting computation path. The construction of these clauses is easy because of the local nature of the computation of a Turing machine. For complete details see Garey and Johnson [37, pages 39–44].

Now $\varphi_n(x, z)$ is true if and only if $z$ corresponds to an accepting computation path. Choose $c$ such that $n^c$ is a bound on both the number of variables and the number of clauses in $\varphi_n$ for all $n$. By Lemma 3.1.1 there exists a polynomial time computable arithmetic formula for $P_n$ that interpolates $\varphi_n$. Thus by summing over all $z$, *i.e.*, over invalid, rejecting and accepting computations, we get that the sum is equal to the number of accepting computations. ∎

The following result is essentially due to J. Simon [68]; similar proofs appear in Peterson and Reif [61] and Orponen [57]. It is a *NEXP* version of the Cook-Levin Theorem. It states that given a language $L$ in *NEXP* there is for each input $x$ a 3-CNF formula with an exponential number of variables and clauses that is satisfiable if and only if $x \in L$. It further states that this 3-CNF formula is simple, in the sense that the $i$th clause is computable in polynomial time.

**Lemma 4.2.2** *Let $L \in NEXP$. Then there exists a constant $c$ such that for every $x \in \{0,1\}^n$ there exists a 3-CNF formula with an exponential number of clauses and an exponential number of variables $X(0), X(1), \ldots, X(2^{n^c} - 1)$:*

$$\Phi_x(X(0), X(1), \ldots, X(2^{n^c} - 1)) = \bigwedge_{i=0}^{2^{n^c}-1} C_i,$$

*where $C_i$ is a clause with three literals. The following two properties hold:*

- *A description of $C_i$ is computable in polynomial time from $x, i$.*

- *For every $x \in \{0,1\}^n$,*

$$x \in L \Longleftrightarrow \exists A : \{0,1\}^{n^c} \to \{0,1\} \text{ such that } \forall i, A \text{ satisfies } C_i.$$

**Proof:** Let $M$ be a *NEXP* Turing machine accepting $L$. Look at the tableau describing the computation of $M$ on input $x$. Convert this to a 3-CNF formula $\Phi_x$ as in the proof of the Cook-Levin theorem. There are an exponential number $N$ of variables and an exponential number $N'$ of clauses. For sake of simplicity, assume without loss of generality that $N = N' = 2^{n^c}$ for some constant $c$. However, the clauses are easily definable, meaning that the type and the indexes of the variable in the $i$th clause are computable in polynomial time. $\blacksquare$

We state the above result in another way, that will facilitate an arithmetization.

Given a clause $C$ with 3 literals $z_1, z_2, z_3$ we define for $j \in \{1,2,3\}$:

$$t_j := \begin{cases} 1 & \text{if } z_j \text{ is not negated in } C \\ 0 & \text{otherwise.} \end{cases}$$

We call $t = (t_0, t_1, t_2)$ the *type* of $C$. And given a type $t$ we define $C_t$ as the unique clause of type $t$ (with variables $z_1, z_2, z_3$).

**Lemma 4.2.3** *Let $L \in NEXP$ and $x \in \{0,1\}^n$. Then there exists a constant $c$ and a polynomial time computable Boolean function $f_x : \{0,1\}^{4n^c+3} \to \{0,1\}$ such that*

$$x \in L \iff \exists A : \{0,1\}^{n^c} \to \{0,1\} : \forall i, b_1, b_2, b_3, t :$$
$$[f_x(i, b_1, b_2, b_3, t) = 1] \Rightarrow [A(b_1), A(b_2), A(b_3) \text{ satisfy } C_t],$$

*where $i, b_1, b_2, b_3 \in \{0,1\}^{n^c}$ and $t \in \{0,1\}^3$.*

**Proof:** Let $f_x(i, b_1, b_2, b_3, t)$ be the Boolean function that is 1 if and only if $b_1, b_2, b_3$ are the indexes of the variables and $t$ is the type of the $i$th clause. The statement that $A$ satisfies the $i$th clause is equivalent to the statement

$$\forall b_1, b_2, b_3, t : [f_x(i, b_1, b_2, b_3, t) = 1] \Rightarrow [A(b_1), A(b_2), A(b_3) \text{ satisfy } C_t].$$

Hence the lemma follows directly from Lemma 4.2.2. ∎

For the rest of this section let us fix a $L \in NEXP$ and a Boolean function $A$. We need to arithmetize the statement $\mathcal{S}_1$:

$$\forall i, b_1, b_2, b_3, t : [f_x(i, b_1, b_2, b_3, t) = 1] \Rightarrow [A(b_1), A(b_2), A(b_3) \text{ satisfy } C_t].$$

By De Morgan's Laws and the definition of implication, this statement is equivalent to the following statement $\mathcal{S}_2$:

$$\neg \exists i, b_1, b_2, b_3, t : [f_x(i, b_1, b_2, b_3, t) = 1] \wedge \neg [A(b_1), A(b_2), A(b_3) \text{ satisfy } C_t].$$

We start by arithmetizing a substatement of $\mathcal{S}_2$:

$$\mathcal{S}_3 := [f_x(i, b_1, b_2, b_3, t) = 1] \wedge \neg [A(b_1), A(b_2), A(b_3) \text{ satisfy } C_t].$$

We let "true" be 1 and "false" be 0.

**Lemma 4.2.4** *There exists a constant $c'$ depending only on $L$ such that for all $x$ there exists a polynomial time computable arithmetic formula $P_x(i, b_1, b_2, b_3, t, z, a)$, with $n^c + 3n^c + 3 + n^{c'} + 3$ variables and degree at most $n^{c'}$ such that for all $i, b_1, b_2, b_3 \in \{0,1\}^{n^c}$ and all $t \in \{0,1\}^3$:*

$$\sum_{z \in \{0,1\}^{n^{c'}}} P_x(i, b_1, b_2, b_3, t, z, A(b_1), A(b_2), A(b_3)) = \begin{cases} 1 & \text{if } \mathcal{S}_3 \text{ is true} \\ 0 & \text{otherwise.} \end{cases}$$

**Proof:** From Lemma 4.2.3 we obtain that $f_x$ is computable in polynomial time. Hence from Lemma 4.2.1 we get a polynomial time computable arithmetic formula $F_x$ and a constant $c'$ such that

$$\sum_{z \in \{0,1\}^{n^{c'}}} F_x(i, t, b_1, b_2, b_3, z) = \begin{cases} 1 & \text{if } f_x(i, t, b_1, b_2, b_3) \text{ is true} \\ 0 & \text{otherwise.} \end{cases}$$

Observe that since $f_x$ is a deterministic polynomial time computable function there is only one valid computation $z$, which either is an accepting or rejecting computation. To arithmetize

$$[A(b_1), A(b_2), A(b_3) \text{ satisfy } C_t]$$

we define the predicate $\rho : \{0,1\}^6 \rightarrow \{0,1\}$ such that

$$\rho(t_1, t_2, t_3, a_1, a_2, a_3) := $$
$$\begin{cases} 1 & \text{if } (a_1, a_2, a_3) \text{ satisfies the clause of type } (t_1, t_2, t_3) \\ 0 & \text{otherwise.} \end{cases}$$

From Proposition 2.1.1 we know that there exists a polynomial $q(t_1, t_2, t_3, a_1, a_2, a_3)$ such that $q$ interpolates $\rho$.

Let

$$P_x(i, b_1, b_2, b_3, t, z, a_1, a_2, a_3) := $$
$$F_x(i, b_1, b_2, b_3, t, z)(1 - q(t_1, t_2, t_3, a_1, a_2, a_3)).$$

Now look at

$$\sum_{z \in \{0,1\}^{n^{c'}}} P_x(i, b_1, b_2, b_3, t, z, A(b_1), A(b_2), A(b_3))$$
$$= \sum_{z \in \{0,1\}^{n^{c'}}} F_x(i, b_1, b_2, b_3, t, z)(1 - q(t_1, t_2, t_3, A(b_1), A(b_2), A(b_3)))$$
$$= \begin{cases} 1 & \text{if } \mathcal{S}_3 \text{ is true} \\ 0 & \text{otherwise} \end{cases} \blacksquare$$

To arithmetize $\mathcal{S}_2$ we will use a different interpretation of the truth values. We let "true" be 0 and "false" be $> 0$.

**Lemma 4.2.5** *There exists a constant $c'$ depending only on $L$ such that for all $x$ there exists a polynomial time computable arithmetic formula $P_x(i, b_1, b_2, b_3, t, z, a)$, with $n^c + 3n^c + 3 + n^{c'} + 3$ variables and degree at most $n^{c'}$ such that*

$$\sum_{i, b_1, b_2, b_3, t, z} P_x(i, b_1, b_2, b_3, t, z, A(b_1), A(b_2), A(b_3)) = \begin{cases} 0 & \text{if } \mathcal{S}_1 \text{ is true} \\ > 0 & \text{otherwise.} \end{cases}$$

*where $i, b_1, b_2, b_3, t, z \in \{0,1\}^{n^{4c+3+c'}}$.*

**Proof:**  This is clear from Lemma 4.2.4, since

$$\sum_{i,b_1,b_2,b_3,t,z} P_x(i, b_1, b_2, b_3, t, z, A(b_1), A(b_2), A(b_3))$$

$$= \sum_{i,b_1,b_2,b_3,t} \sum_z P_x(i, b_1, b_2, b_3, t, z, A(b_1), A(b_2), A(b_3))$$

$$= \sum_{i,b_1,b_2,b_3,t} \begin{cases} 1 & \text{if } \mathcal{S}_3 \text{ is true} \\ 0 & \text{otherwise.} \end{cases}$$

$$= \begin{cases} 0 & \text{if } \mathcal{S}_2 \text{ is true} \\ > 0 & \text{otherwise.} \end{cases} \blacksquare$$

This gives us the arithmetization of Lemma 4.2.3, which we need to construct a multiple-prover interactive proof system for every language in *NEXP*.

**Lemma 4.2.6** *Let* $L \in NEXP$. *Then there exist constants $c$ and $c'$ such that for all $n$ and for all $x \in \{0,1\}^n$ there exists an arithmetic formula $P_x(i, b_1, b_2, b_3, t, z, a)$, with $n^c + 3n^c + 3 + n^{c'} + 3$ variables and degree at most $n^{c'}$, such that*

$$x \in L \iff \exists A : \{0,1\}^{n^c} \to \{0,1\} \text{ such that}$$
$$\sum_{i,b_1,b_2,b_3,t,z} P_x(i, b_1, b_2, b_3, t, z, A(b_1), A(b_2), A(b_3)) = 0,$$

*where* $i, b_1, b_2, b_3, t, z \in \{0,1\}^{n^{4c+3+c'}}$.

*Furthermore $P_x$ is computable in polynomial time.*

## 4.3  Multiple Prover Interactive Proofs for *NEXP*

In this section, we will describe how we apply the technique from single-prover interactive proof systems to the arithmetization of *NEXP*.

Let $L$ be a language in *NEXP*.

From the alternative characterization of multiple-prover interactive proof systems we know that the provers can provide an oracle that tells the satisfying assignment $A$. So the only thing the verifier has to verify is that

$$\sum_{i,t,b_1,b_2,b_3,z} P_x(i, b_1, b_2, b_3, t, z, A(b_1), A(b_2), A(b_3)) = 0.$$

In the proof systems for $\#P$ the verifier had to verify the value of a similar expression. The property used for $\#P$ was that the summand was a low degree polynomial. Here $P_x(i, b_1, b_2, b_3, t, z, A(b_1), A(b_2), A(b_3))$ is not a polynomial, because of the Boolean function $A$. To insure that we have a polynomial we extend the domain of $A$ to all $\mathbf{Z}^{n^c}$. We do this by using Proposition 2.1.1 that tells us that there exists a unique multilinear polynomial $p_A$ that interpolates $A$. The verifier has to check that

$$\sum_{i,t,b_1,b_2,b_3} P_x(i, b_1, b_2, b_3, t, z, p_A(b_1), p_A(b_2), p_A(b_3)) = 0. \qquad (4.1)$$

Since $P_x(i, b_1, b_2, b_3, t, z, p_A(b_1), p_A(b_2), p_A(b_3))$ is a polynomial and each of its variables has low degree we can use the techniques from the single-prover interactive proof to verify 4.1. In the verification of Equation 4.1 the verifier asks for values $p_A(\alpha)$ for some $\alpha \in \{0, 1, \ldots, N-1\}^{n^c}$, where $N$ is a parameter chosen to make that probability of error in the verification suitable small. We choose $N = 6(n^{c'}+3)(4n^c+n^{c'}+3)$. The oracle the provers construct, therefore, contains the values of $p_A(\alpha)$ for all $\alpha \in \{0, 1, \ldots, N-1\}^{n^c}$. We call the oracle $A'$. If the verifier accepts that

$$\sum_{i,t,b_1,b_2,b_3} P_x(i, b_1, b_2, b_3, t, z, A'(b_1), A'(b_2), A'(b_3)) = 0.$$

then it knows with high probability that $x \in L$ *or* that the provers did not let $A'$ be a multilinear polynomial that interpolates a Boolean function. So if $x \notin L$ and the verifier accepts with high probability then either

1. $A'$ is not multilinear, or

2. $A'(b) \notin \{0,1\}$ for some $b \in \{0,1\}^{n^c}$.

We shall discuss the verification of (approximate) multilinearity later, but first, observe that the verifier can guard against the second possibility by verifying that

$$\sum_{b \in \{0,1\}^{n^c}} (A'(b)(1 - A'(b)))^2 = 0, \tag{4.2}$$

in the following way.

Note that (4.2) will be the case if and only if

$$\forall b \in \{0,1\}^{n^c} : A'(b) \in \{0,1\}. \tag{4.3}$$

If we assume that $A'$ is multilinear then the verifier can verify 4.2 using the single-prover techniques. Hence, the verifier, with the help of one of the provers, eliminates the summation and ends up by having to evaluate a simple formula. Note that in this part it is crucial that we use the ring of integers and not a finite field, since if we were to use a finite field, Equation 4.2 would not be equivalent to Equation 4.3. There exists a way to overcome this problem and use finite fields anyway. This can be found in [9].

Using the integers leaves us with one problem not encountered in using finite fields. The proof that the verifier only uses polynomial time is more complicated. If we use finite fields, we only needed to count the number of arithmetic operations, whereas, if we use integers we also need to show that the numbers do not become too large. In fact, the protocol used to show that every language in *PSPACE* has a single-prover interactive proof (see section 3.2) would have required more than polynomial time if instead of finite fields we had used integers. This is because the coefficients of some of the polynomials would have had doubly exponential absolute value.

Because of these complications we give all the details for the verification of Equation 4.2. The protocol uses the same ideas as the protocols in Chapter 3 (See Figure 4.) The verification of Equation 4.1 is similar.

**Lemma 4.3.1** *If $A'$ is multilinear then Protocol 4 satisfies the following statements:*

1. *If $\sum_b (A'(b)(1 - A'(b)))^2 = 0$ then there exists a prover such that the verifier always accepts.*

2. *If $\sum_b (A'(b)(1 - A'(b)))^2 \neq 0$ then for all provers, the verifier accepts with probability at most $\frac{4n^c}{N}$.*

---

**Protocol 4**

**V:** $N \leftarrow 6(n^{c'} + 3)(4n^c + n^{c'} + 3)$.

**V:** $\beta_0 \leftarrow 0$.

    **for** $i = 1$ **to** $n^c$ **do**

        **P→V:** $\gamma_0, \gamma_1, \gamma_2, \gamma_3, \gamma_4$ integers whose absolute value is bounded by $2^{9n^c} N^{4n^c}$ .

        **V:** $q_i(x) \leftarrow \sum_{j=0}^4 \gamma_j x^j$.

        **V:** Check that $q_i(0) + q_i(1) = \beta_{i-1}$ **else** halt and reject.

        **V:** Choose $\alpha_i$ uniformly and randomly in $\{0, 1, \ldots, N-1\}$.

        **V:** $\beta_i \leftarrow q_i(\alpha_i)$.

    **end**

**V: if** $\beta_{n^c} = (A'(\alpha_1, \alpha_2, \ldots, \alpha_{n^c}))(1 - A'(\alpha_1, \alpha_2, \ldots, \alpha_{n^c}))^2$ **then**

    **Halt** and **Accept**

**else**

    **Halt** and **Reject**

---

Figure 4.1: Protocol for verification that $\sum_b (A'(b)(1 - A'(b)))^2 = 0$.

*3. The verifier runs in polynomial time.*

**Proof:**

1. This is clear except we have to show that in each round the honest prover can provide the verifier with coefficients of absolute value less than $2^{9n^c} N^{4n^c}$. Consider the $i$th round. The polynomial that the prover is supposed to give the verifier is

$$\sum_{b \in \{0,1\}^{m-i}} (A'(\alpha, x, b)(1 - A'(\alpha, x, b)))^2, \tag{4.4}$$

where $\alpha \in \{0, 1, \ldots, N-1\}^{i-1}$ and $m = n^c$. So we have to bound the coefficients of this polynomial.

Let us first find a bound on the absolute value of the polynomial in $x$:

$$A'(\alpha_1, \alpha_2, \ldots, \alpha_{i-1}, x, b_{i+1}, b_{i+2}, \ldots, b_m),$$

where $\alpha_1, \alpha_2, \ldots, \alpha_{i-1} \in \{0, \ldots, N-1\}$ and $b_{i+1}, b_{i+2}, \ldots, b_m \in \{0, 1\}$.

We know that $A'(b) \in \{0, 1\}$ for all $b \in \{0, 1\}^m$, and we know from the proof of Proposition 2.1.1 that

$$A'(x_1, x_2, \ldots, x_m) = \sum_{z \in \{0,1\}^m} A'(z) \prod_{j=1}^{m} (1 - z_j - x_j + 2x_j z_j).$$

Recall that for $x_j, z_j \in \{0, 1\}$, $1 - z_j - x_j + 2x_j z_j$ is 1 if and only if $x_j = z_j$, and 0 otherwise.

Thus the absolute values of the coefficients of $A'(\alpha, x, b)$ are, at most, $2^m (2N)^i$. Hence the polynomial 4.4 has coefficients of absolute value, at most, $2^{m-i-1} (2^m (2N)^i)^4 \leq 2^{9m} N^{4m}$.

2. This is clear since the probability of failure in each round is at most $\frac{4}{N}$, since $q_i(x)$ is a polynomial of degree at most 4.

3. Follows since the number of arithmetic operations is polynomial in $n$ and since the integers used only have a polynomial number of digits.  ∎

A similar proof shows that

**Lemma 4.3.2** *If $A'$ is multilinear then a protocol similar to Protocol 4 satisfies the following statements:*

1. *If Equation 4.1 is true then there exists a prover such that the verifier always accepts.*

2. *If Equation 4.1 is false then for all provers, the verifier accepts with probability at most $\frac{(n^{c'}+3)(4n^c + n^{c'}+3)}{N} = \frac{1}{6}$.*

3. *The verifier runs in polynomial time.*

The only remaining problem is that the provers can try to gain an advantage by having the function $A$ not be multilinear. Indeed, $A$ could be some horribly complex functions intended to foul up the above protocol that uses heavily the fact that $A$ is a low-degree polynomial.

Note that there is no probabilistic polynomial time test that can test if $A'$ is truly multilinear since there are exponential number of points in the domain of $A'$. However, we show that there is a probabilistic polynomial time test that with high probability discovers if $A'$ is not "close" to being multilinear. We say that $A'$ is *$\delta$-approximately multilinear* if there exists a multilinear function $g$ such that $g(x) = A(x)$ for at least $(1-\delta)$ fraction of all $x \in \{0, 1, \ldots, N-1\}^n$. The following result, which will be proven in the next section, gives us the multilinearity test.

**Lemma 4.3.3** *Let $N$ be an integer, $40n^2 < N \leq 2^n$ for some $n$. Let $\mathbf{I}$ denote the set of integers $\{0, \ldots, N-1\}$. Let $A(x_1, \ldots, x_n)$ be an arbitrary function from $\mathbf{I}^n$ to $\mathbf{Q}$. Then there exists a probabilistic polynomial time Turing machine $M$ such that, given access to an oracle $A$,*

1. *If $A$ is multilinear, integral valued, and does not take absolute values greater than $2^n N^n$ then $M^A$ always accepts.*

2. *If $A$ is not $\frac{1}{24}$-approximately multilinear then $M^A$ rejects with probability at least $\frac{1}{3}$.*

The whole protocol is this: the verifier runs the approximate multilinearity test from Lemma 4.3.3 and verifies Equation 4.1 and Equation 4.2 (See Figure 4.2).

**Theorem 4.3.4** *Protocol 5 satisfies the following statements:*

1. *If $x \in L$ then there exists a pair of provers such that the verifier always accepts.*

2. *If $x \notin L$ then for all pairs of provers, the verifier accepts with probability at most $\frac{1}{3}$.*

3. *The verifier runs in polynomial time.*

---

**Protocol 5**

$\mathbf{P_1, P_2}$: Provides an oracle for $A' : \mathbf{I}^n \to \mathbf{Z}$. In section 4.3.1 we discuss how this is done.

  $\mathbf{V}$: Verify that $A'$ is approximately multilinear using Lemma 4.3.3

  $\mathbf{V}$: Verify that

$$\sum_b (A'(b)(1 - A'(b)))^2 = 0,$$

using Protocol 4 and $P_1$ as the one prover needed in that protocol.

  $\mathbf{V}$: Verify that

$$\sum_{i,t,b_1,b_2,b_3} P_x(i, b_1, b_2, b_3, t, z, A'(b_1), A'(b_2), A'(b_3)) = 0,$$

using a protocol similar to Protocol 4.

---

Figure 4.2: Protocol for $L \in NEXP$.

**Proof:**

1. This is obvious from the arithmetization (Lemma 4.2.6), the first part of the proof of Lemma 4.3.1 and the above discussion.

2. We know from Lemma 4.3.3 that if $A'$ is not $\frac{1}{24}$-approximately multilinear then the verifier accepts with probability at most $\frac{1}{3}$. Therefore, we can assume that the oracle is $\frac{1}{24}$-approximately multilinear and that there exists a multilinear polynomial $g$ that agrees with $A'$ for at least a $\frac{23}{24}$ fraction of $\mathbf{I}^{n^c}$. In the verification of 4.1 and 4.2 the verifier only asks 4 questions of $A'$. Thus there is only a $\frac{1}{24}$ chance for each question that $g$ and $A'$ do not agree, since the questions to $A'$ are chosen randomly and uniformly in $\mathbf{I}^{n^c}$. So with probability at least $\left(\frac{23}{24}\right)^4 > \frac{5}{6}$, $g$ and $A'$ agree on the 4 questions. In which case, the verifier accepts with oracle $g$ if and only if it accepts with oracle $A'$. But since $g$ is multilinear we know from Lemma 4.3.1 and Lemma 4.3.2 that the verifier accepts a false statement about Equations 4.1 or 4.2 with probability at most $\frac{1}{6}$. But since $x \notin L$ we know that the verifier must accept at least one false statement.

Let us repeat the argument. Let us denote:

$I_1 :=$ "the verifier accepts with $A'$."

$I_2 :=$ "$A'$ is not $\frac{1}{24}$-approximately multilinear, but the verifier accepts."

$I_3 :=$ "the verifier asks one question on which $g$ and $A'$ disagree."

$I_4 :=$ "the verifier accepts with $g$."

So

$$\begin{aligned}
\Pr[\text{verifier accepts with } A'] &= \Pr[I_1] \\
&\leq \max(\Pr[I_2], \Pr[I_3] + \Pr[I_4]) \\
&\leq \max(\frac{1}{3}, 1 - \left(\frac{23}{24}\right)^4 + \frac{1}{6}) \\
&= \frac{1}{3}. \blacksquare
\end{aligned}$$

This completes the proof of Theorem 4.3.4 modulo the proof of Lemma 4.3.3. Let us first discuss an easy two-prover implementation of the protocol that refers to $A$ as an oracle (predetermined function).

## 4.3.1 Implementing the Protocol with Two Provers

We show, in this section, how to use two provers to execute the above protocol based on the work of Fortnow, Rompel and Sipser and Ben-Or, Goldwasser, Kilian and Wigderson described in Section 4.1. We ask prover 2 exactly one question about $A'$. In fact, we define $A'$ as the function of prover 2's responses to the single question it is asked.

We execute the entire protocol with prover 1 answering questions about $A'$. We then choose a random question we have asked prover 1 about $A'$ and ask this question to prover 2. If the answers differ then we reject.

If the provers have been honest then all questions about $A'$ have the same answer with prover 1 and prover 2. If prover 1 gives an incorrect answer about $A'$ then the verifier catches him with probability at least $1/n^k$ where $n^k$ is greater then the number of questions about $A'$ that the verifier has asked of prover 1.

We repeat the above protocol $n^{k+1}$ times. If prover 1 has to lie about some value of $A'$ on each run of the protocol then the verifier catches the prover with probability $> 1 - e^{-n}$.

### 4.3.2   The Power of the Provers

As a byproduct of the above protocol, we get the following corollary regarding the required power of the provers:

**Corollary 4.3.5** *For any $L \in EXP$, there is a multiple-prover interactive proof system with provers living in EXP.*

**Proof:** Notice that the tableau of the computation performed by a deterministic exponential-time machine $M$ on a specific input $x$ is unique and that any bit of the tableau can be computed in deterministic exponential time by simulating the computation of $M(x)$. The representation of an exponential-time function is still exponential time by interpolating the exponential number of points needed to specify the representation. From this it is easy to see that for the honest provers, the power of deterministic exponential time suffices. ∎

Although we know all languages provable in a multiple-prover proof system must lie in $NEXP$, we do not know whether $NEXP$ provers are sufficient to prove any $NEXP$ language to a verifier. It would be important that all provers have access to the same tableau of an accepting computation; but, there could be several since the $NEXP$ machine may have many accepting paths. The best upper bound we know is $EXP^{NP}$ provers that can identify the lexicographically first tableau.

## 4.4   Verification of Multilinear Functions

In this section we show how to test a function for approximate multilinearity. We prove a more general theorem than Lemma 4.3.3, the one used in the preceding section. We prove the following:

**Theorem 4.4.1** *Let $\alpha$ and $p$ be positive reals and let $n, K$ and $N$ be integers such that $\frac{40n^2}{\alpha} < N$. Let $\mathbf{I}$ denote the set of integers $\{0, \ldots, N-1\}$. Let $A(x_1, \ldots, x_n)$ be an arbitrary function from $\mathbf{I}^n$ to $\mathbf{Q}$. Then there exists a probabilistic Turing machine $M$, which works in time in*

$$O\left(n^5 \frac{1}{\alpha'^2}(\log^2 K)\log^2 \frac{1}{p}\right),$$

*where $\alpha' = \min\left\{\alpha, \frac{1}{n}\right\}$, such that given access to $A$ as an oracle:*

- *If $A$ is multilinear, integral valued, and does not take absolute values greater than $K$ then $M^A$ always accepts.*

- *If $A$ is not $\alpha$-approximately multilinear then $M^A$ rejects with probability at least $1 - p$.*

**Remark 4.4.2** Recently, other tests for multilinearity/low-degreeness have been designed [26, 4, 3], that are more efficient and in some case have a simpler proof of correctness. For the simplest test and proof see [26].

The rest of this section is dedicated to proving this theorem. We discuss further generalizations of Theorem 4.4.1 in section 4.4.7.

First we need some definitions and notation.

We shall consider the $n$th Cartesian power of a finite set $X$ (usually $X = \mathbf{I}$). A subset $U \subseteq X^n$ is called a *$k$-dimensional subspace* of $X^n$ if there exist $n - k$ different coordinates $i_1, \ldots, i_{n-k}$ and $n - k$ values of these coordinates $\alpha_{i_1}, \ldots, \alpha_{i_{n-k}} \in X$ such that

$$U = \{(\alpha_1, \ldots, \alpha_n) : x_{i_j} = \alpha_{i_j} \text{ for } j = 1, 2, \ldots, n-k\}.$$

In other words a $k$-dimensional subspace is a set of points where $n - k$ coordinates are fixed and the rest take on all possible values.

A *line* is a 1-dimensional subspace. The points of a line in the $k$th *direction* have all but the $k$th coordinate in common. We shall denote by $L$ the set of lines and by $L_i$ the set of $|X|^{n-1}$ lines in the $i$th direction. A *hyperplane* is a subspace of dimension $n - 1$. We shall use these terms for the case $X = \mathbf{I}$. Note that what we call subspaces correspond to the subspaces *aligned* with the coordinate system in the affine space. (In this terminology, $\mathbf{I}^n$ has $nN$ hyperplanes.) We define the natural *measure* of a subset $Y$ of a finite set $X$ by $\mu_X(Y) := \frac{|Y|}{|X|}$. We normally drop the subscript and just let $u(Y)$ denote the measure of $Y$ if $X$ is obvious from the context.

**Definition 4.4.3** *Let $f, g$ be functions over a finite set $X$. For $\delta \in [0, 1]$ we say that $f$ $\delta$-approximates $g$ if $\mu(\{x \in X | f(x) \neq g(x)\}) < \delta$.*

**Definition 4.4.4** *Let $f : \mathbf{I}^n \to \mathbf{Q}$ be a function. We call $f$ multilinear if its restriction to any line (in the above sense) of $\mathbf{I}^n$ is linear.*

**Definition 4.4.5** *Let $f : \mathbf{I}^n \to \mathbf{Q}$. For $\delta \in [0, 1]$ we say that $f$ is $\delta$-approximately multilinear if there exists a multilinear $g$ such that $g$ $\delta$-approximates $f$. If n=1, we obtain the concept of $\delta$-approximately linear functions.*

**Definition 4.4.6** *Given a function $A : \mathbf{I}^n \to \mathbf{Q}$, we call a line $\ell$ in $\mathbf{I}^n$ correct if the restriction $A|_\ell$ is a linear function. We say that $\ell$ is $\delta$-wrong or just wrong if $A|_\ell$ is not $\delta$-approximately linear.*

## 4.4.1   The Test for Multilinearity

The test is simple. It takes its main idea just from the definition of multilinearity (see Definition 4.4.4). For each variable we pick randomly and uniformly many lines and test that on each of the picked lines $A$, restricted to the line, is a linear function. After this, we know with high probability that $A$, restricted to most lines, is a linear function. This, we later show is enough to be sure that $A$ is approximately multilinear. One problem with this is that we can not even test, given a line $l$, if $A$ restricted to $l$ is linear since we may have an exponential number of points on $l$. To test that $A$ restricted to $l$ is a linear function we first find the linear function $f$ such that $f(0) = A|_l(0)$ and $f(1) = A|_l(1)$ and then, we randomly and uniformly pick many points on the line and verify that on those point $A_l$ is equal to $f$. Furthermore we reject $A$ if we ever finds an absolute value of $A$ that is larger than $K$ or not integral. For a complete description of the test see Figure 4.3.

```
for i := 1 to n do {Test direction i}
    for j := 1 to m₁ do {Pick a random line l.}
        Randomly and uniformly pick α₁, .., αᵢ₋₁, αᵢ₊₁, .., αₙ ∈ I.
        {Find f(x) = ax + b, such that f(0) = A|ₗ(0) and f(1) = A|ₗ(1).}
        a ← A(α₁, .., αᵢ₋₁, 1, αᵢ₊₁, .., αₙ) − A(α₁, .., αᵢ₋₁, 0, αᵢ₊₁, .., αₙ).
        b ← A(α₁, .., αᵢ₋₁, 0, αᵢ₊₁, .., αₙ).
        for k := 1 to m₂ do {Check a random point on l.}
            Randomly and uniformly pick βₖ ∈ I.
            γₖ ← A(α₁, .., αᵢ₋₁, βₖ, αᵢ₊₁, .., αₙ)
            if γₖ ≠ aβₖ + b or |γₖ| > K or γₖ is not integral then
                Halt and Reject
        end
    end
end
Halt and Accept
```

Figure 4.3: The multilinearity test.

**Proposition 4.4.7** *Given a function $A : \mathbf{I}^n \rightarrow \mathbf{Q}$, assume that $A$ passes the test with $m_1 = \lceil (\ln \frac{2}{p})/\epsilon \rceil, m_2 = \lceil (\ln \frac{2}{p}/\delta) \rceil$. Then we infer with confidence $\geq 1 - p$ that*

$$(\forall i) \text{ the proportion of } \delta\text{-wrong lines among } L_i \text{ is } < \epsilon. \qquad (4.5)$$

**Proof:** Assume that the proportion $\rho$ of wrong lines among $L_i$ is larger than $\epsilon$ for some direction $i$. We have to show that then the test accepts $A$ with probability at most $p$. There are two ways that the test can fail.

- It does not choose a wrong line.

- It chooses a wrong line, but does not discover that it is wrong.

The probability of the first event is bounded by the probability that the test does not choose a wrong line in direction $i$. The probability of not choosing a wrong line in direction $i$ is $(1 - \rho)^{m_1} \leq (1 - \epsilon)^{m_1} < e^{-\epsilon m_1}$.

The probability of the second event is similarly bounded. Given a wrong line $l$ let $f : \mathbf{I} \rightarrow \mathbf{Q}$ be the linear function defined by $f(0) = A|_l(0)$ and

$f(1) = A|_l(1)$. Let $B \subset l$ be the set of points where $f$ and $A|_l$ do not agree. Since $l$ was wrong we know that $\mu(B) \geq \delta$. The probability that we do not pick a point in $B$ is, therefore, $(1 - \mu(B))^{m_2} \leq (1 - \delta)^{m_2} < e^{-\delta m_2}$.

The probability that the test fails is bounded by $e^{-\epsilon m_1} + e^{-\delta m_2}$, which is equal to $p$ by the choice of $m_1$ and $m_2$. ∎

The rest of this section is devoted to proving that the above conclusion 4.5 implies that $A$ is $\epsilon'$-approximately multilinear for some small $\epsilon'$. See Theorem 4.4.13, at the end of this section, for the formal statement of this result. We prove Theorem 4.4.13 by proving a series of four lemmas. The whole argument is an inductive argument that constructs a multilinear approximation to $A$. The inductive argument is non-standard and it is proven in the Tree-coloring Lemma. The inductive step is proven in the Pasting Lemma, that uses the Self-Improvement Lemma, which improves the quality of the approximation constructed in the inductive step. To prove the Self-Improvement Lemma we need a combinatorial isoperimetric inequality, which is proven in the Expansion Lemma.

## 4.4.2    The Expansion Lemma

**Definition 4.4.8** *Let $X$ be a finite set. For $S \subseteq X^n$ define the* neighbors *of $S$ as*

$$\bar{S} = \bigcup_{i=1}^{n} \pi_i^{-1}(\pi_i(S)),$$

*where $\pi_i : X^n \to X^{n-1}$ is the projection in the ith direction.*

In other words, the neighbors $\bar{S}$ of a set $S$ is the set of points $x$ such that there exists a point $y$ in $S$ that only differs from $x$ in at most one coordinate.

**Lemma 4.4.9 (Expansion Lemma)** *Let $S \subseteq X^n$. If $|S| \leq |X|^n/2$ then $|\bar{S}| \geq |S|(1 + \frac{1}{2n})$.*

**Proof:**   This was proved by Aldous [2, Lemma 3.1]. It is also implicit in work by Babai and Erdős [6, Lemma]. For completeness we include it here.

We look at $X^n$ as $\mathbf{Z}_N^n$, where $N = |X|$. For $x \in \mathbf{Z}_N^n$ define

$$Sx := \{y + x \mid y \in S\}.$$

Then, observe that

$$\sum_{x \in \mathbf{Z}_N^n} |S \cap Sx| = |S|^2.$$

Hence there exists a $x' \in \mathbf{Z}_N^n$ such that $|S \cap Sx'| \leq |S|^2/N^n$. This implies that

$$|Sx' \setminus S| \geq |S| - |S|^2/N^n \geq \frac{1}{2}|S|.$$

Now write $x' = y_1 + y_2 + \cdots + y_n$, where $y_i = (0, \ldots, 0, x_i', 0, \ldots, 0)$ and let $z_i := y_1 + y_2 + \cdots + y_i$. Then

$$|Sx' \setminus S| \leq \sum_{i=1}^n |Sz_i \setminus Sz_{i-1}| = \sum_{i=1}^n |Sy_i \setminus S|.$$

Thus there exists an $i$ such that $|Sy_i \setminus S| \geq \frac{1}{2n}|S|$ and therefore $|\bar{S} - S| \geq |Sy_i \setminus S| \geq \frac{1}{2n}|S|$. $\blacksquare$

## 4.4.3 The Self-Improvement Lemma

The key step in the induction argument that yields Theorem 4.4.13 is the verification that if a function passes the test and it is multilinear on a fair portion of the space then it is really a good approximation to a multilinear function. Here is the formal statement:

**Lemma 4.4.10 (Self-improvement lemma)** *Given a function* $A : \mathbf{I}^n \to \mathbf{Q}$, *assume that*

$(\forall i)$ *the proportion of wrong lines among* $L_i$ *is* $< \epsilon$

*and*

$\exists g : g$ *is multilinear and* $g$ $\Delta$-*approximates* $A$,

*where* $\Delta \leq 1/2$. *Then*

$g$ $\epsilon'$-*approximates* $A$, *where* $\epsilon' = 3n^2(\epsilon + \delta + 1/N)$.

**Proof:**   We partition the points of $\mathbf{I}^n$ into four sets.

$B$: $B = \{x \in \mathbf{I}^n | A(x) \neq g(x)\}$. Call $x \in B$ a bad point.

$W$: Union of points on wrong lines. Call the points in $W$ wrong. Observe that the first assumption gives $\mu(W) < n\epsilon$.

$M$: Points $(x \notin W)$ which belong to lines $\ell$ where $A|_\ell$ is $\delta$-approximated by some linear function $h$, but $h(x) \neq A(x)$. Call these points *misplaced*. Since for each line only a $\delta$-fraction is misplaced and since each point lies on $n$ lines we obtain that $\mu(M) < n\delta$.

$I$: Points $x$ on lines $\ell$ such that $A|_\ell$ is $\delta$-approximated by a linear function $h$, where $h \neq g|_\ell$, but $A(x) = g(x) = h(x)$. Since, at most, one such point belongs to each line, $\mu(I) \leq n/N$.

Let us define $S := B \setminus (W \cup M)$. We claim that $\bar{S} \subseteq B \cup M \cup I$. To see this take $\alpha \in S$ and let $\beta$ be a point on a line $\ell$ through $\alpha$. Assume that $\beta \notin B \cup M$. Observe that since $\alpha$ is not a wrong point there is a linear function $h$ which $\delta$-approximates $A$ restricted to $\ell$. Since $\alpha \in B \setminus M$ we get that $A(\alpha) = h(\alpha) \neq g(\alpha)$, and hence that $g|_\ell \neq h$. Also since $\beta \notin M$ we get that $h(\beta) = A(\beta)$ and furthermore since $\beta \notin B$ we get that $g(\beta) = A(\beta)$. Thus $\beta \in I$.

If $\mu(S) \leq 1/2$ then from the Expansion Lemma we obtain that

$$(1 + \frac{1}{2n})\mu(S) \leq \mu(\bar{S}) \leq \mu(S) + n(\epsilon + \delta) + n/N,$$

therefore

$$\mu(S) \leq 2n^2(\epsilon + \delta + 1/N).$$

This concludes the proof since

$$
\begin{aligned}
\mu(B) &\leq \mu(S) + \mu(W) + \mu(M) \\
&\leq \mu(S) + n(\epsilon + \delta) \\
&\leq 3n^2(\epsilon + \delta + 1/N). \blacksquare
\end{aligned}
$$

### 4.4.4 The Pasting Lemma

The multilinear function which closely approximates $A$ is constructed for certain subspaces by induction on their dimension. What we show below is that if $A$ is approximately linear on most lines and approximately multilinear on a fair portion of the hyperplanes, then it is approximately multilinear on the entire space. The "self-improvement lemma" prevents the devaluation, through repeated application in the induction argument, of the term "approximately" in this result.

**Lemma 4.4.11 (Pasting Lemma)** *Given a function $A : \mathbf{I}^n \to \mathbf{Q}$, assume that $\delta, \epsilon > 0$, $\epsilon + \delta \le \frac{1}{200n}$, $N \ge 40n$,*

$$(\forall i) \text{ the proportion of } \delta\text{-wrong lines among } L_i \text{ is } < \epsilon$$

*and $\exists g : g(x, y)$ is multilinear in $y \in \mathbf{I}^{n-1}$ and the set*

$$\Phi := \{\xi | g(\xi, y) \ \beta\text{-approximates } A(\xi, y)\}, \text{ has fair density: } \mu(\Phi) \ge \varphi,$$

*where $\varphi = \frac{1}{10n}$ and $\beta = \frac{1}{10}$.*

*Then $A$ is $\Delta$-approximately multilinear, where $\Delta = \epsilon + \delta + 4\beta \le 1/2$. And by the self-improvement lemma $A$ is $\epsilon'$-approximately multilinear, where $\epsilon' = 3n^2(\epsilon + \delta + 1/N)$.*

**Proof:** Observe that the first part of the assumption implies that there exist functions $f_1(y)$ and $f_2(y) : \mathbf{I}^{n-1} \to \mathbf{Q}$ such that $x f_1(y) + f_2(y)$ $(\epsilon + \delta)$-approximates $A(x, y)$.

Define

$$\Psi = \{\xi \mid \xi f_1(y) + f_2(y) \ \beta\text{-approximates } A(\xi, y) \text{ as a function of } y.\}$$

We know that

$$\left| \left\{ (x, y) \mid x \in \mathbf{I}, y \in \mathbf{I}^{n-1} \text{ and } x f_1(y) + f_2(y) \ne A(x, y) \right\} \right| \le (\epsilon + \delta) N^n.$$

For $\xi \in \mathbf{I}$ if $\xi f_1(y) + f_2(y)$ does not $\beta$-approximate $A(\xi, y)$ then the number of $y$s such that $x f_1(y) + f_2(y) \ne A(x, y)$ is at least $\beta N^{n-1}$. Thus

$$\mu(\Psi) N \beta N^{n-1} \le (\epsilon + \delta) N^n.$$

Then

$$\mu(\Psi) \leq \frac{\epsilon + \delta}{\beta} = 10(\epsilon + \delta)$$

and

$$|\Phi \setminus \Psi| \geq N\left(\frac{1}{10n} - 10(\epsilon + \delta)\right) \geq 2,$$

because of our choice of parameters. Let $\xi_1, \xi_2 \in \Phi \setminus \Psi$ such that $\xi_1 \neq \xi_2$. Then for $i = 1, 2$, there exists a multilinear function $g_i(y)$ that $2\beta$-approximate $\xi_i f_1(y) + f_2(y)$. Thus on a set of measure $1 - 4\beta$ we have that

$$f_1(y) = \frac{g_1(y) - g_2(y)}{\xi_1 - \xi_2}$$

and

$$f_2(y) = \frac{\xi_2 g_1(y) - \xi_1 g_2(y)}{\xi_2 - \xi_1}.$$

Denote the multilinear functions on the right hand side by $\tilde{f}_1(y), \tilde{f}_2(y)$. Then, the multilinear function $x\tilde{f}_1(y) + \tilde{f}_2(y)$ $\Delta$-approximates $A(x, y)$. $\blacksquare$

## 4.4.5 The Tree Coloring Lemma

The next lemma provides the overall structure of the induction. It demonstrates, as we shall see in the next subsection, that the Pasting lemma is strong enough to carry approximate multilinearity all the way from most lines to the entire space.

Let $T$ be a depth $n$ levelwise uniform tree (vertices on the same level have the same number of children). We color the tree by two colors red and white. The input is a coloring of the leaves. Then we color the tree bottom up according to the following rule.

Fix the parameters $\epsilon_0$ and $\varphi$. $0 < \epsilon_0, \varphi \leq 1$. Color a vertex red if and only if each of the following two conditions is met:

- *Almost all leaves* in the subtree $T_v$ rooted at $v$ are red: at most an $\epsilon_0$ fraction of $T_v$ leaves are white.

- A *fair number* of children of $v$ are red: the proportion of red children is $\geq \varphi$.

**Lemma 4.4.12 (Tree coloring lemma)** *Let* $\epsilon_k = (1 - \varphi)^k \epsilon_0$. *Let* $v$ *be a vertex on level* $k$. *(The leaves are on level 0.) Assume that all but an* $\epsilon_k$ *fraction of the leaves in* $T_v$ *are red. Then* $v$ *is red.*

**Proof:** By induction on $k$.

$k = 1$ By assumption, the proportion of red children of $v$ is at most $1 - \epsilon_1 = 1 - (1 - \varphi)\epsilon_0 \geq \varphi$ and the proportion of white leaves of $T_v$ is less than $(1 - \varphi)\epsilon_0 \leq \epsilon_0$ so $v$ is red.

$k \geq 2$ Assume that $v$ is colored white. Since $\epsilon_k < \epsilon_0$ we know that the fraction of red children of $v$ is less than $\varphi$. Hence the proportion of white children of $v$ is larger than $1 - \varphi$. For each white child $u$ of $v$ there is, by the inductive hypothesis, at least a fraction of $\epsilon_{k-1}$ white leaves in $T_u$. Therefore the fraction of white leaves in $T_v$ is larger than $(1-\varphi)\epsilon_{k-1} = \epsilon_k$. This contradicts the assumption that there were at most a fraction of $\epsilon_k$ white leaves. $\blacksquare$

## 4.4.6 Conclusion

**Theorem 4.4.13** *Given* $A : \mathbf{I}^n \to \mathbf{Q}$, *assume that*

$$(\forall i) \text{ the proportion of } \delta\text{-wrong lines among } L_i \text{ is } < \epsilon.$$

*Then*

$$A \text{ is } \epsilon'\text{-approximately multilinear, where } \epsilon' = 3n^2(\epsilon + \delta + 1/N),$$

*assuming the parameters have been so chosen that* $N \geq 40n^2$, $\delta \leq \frac{1}{400n^2}$ *and* $\epsilon \leq \frac{1}{800n^3}$.

**Proof:** We first construct a tree $T$ of depth $n - 1$. The nodes of the tree correspond to subspaces of $\mathbf{I}^n$. (We consider aligned affine subspaces; see the conventions stated at the beginning of Section 4.4.1.) The root corresponds to $\mathbf{I}^n$; and the children of a node correspond to its hyperplanes. Observe that the leaves correspond to lines. Note that a line corresponds to $n - 1$ leaves. We color a leaf white if and only if it corresponds to a wrong line. We color the rest of $T$ according to the coloring rule with $\epsilon_0 = 2\epsilon$ and $\varphi = \frac{1}{10n}$. Note that $\epsilon < \epsilon_n = (1 - \varphi)^n \epsilon_0$. From the coloring lemma we obtain that the root is red. Now we only have to make the following observation. We shall say that a subspace $U$ is $\beta$-approximately multilinear if the restriction $A|_U$ is $\beta$-approximately multilinear.

**Lemma 4.4.14** *If $v \in T$ is red then the subspace $U_v$ corresponding to $v$ is $\beta$-approximately multilinear, where $\beta = 1/10$.*

**Proof:** By induction on the level $k$ of $v$.

$k = 1$ Follows since $\delta < \beta$.

$k \geq 1$ We know that since $v$ is red, it has a fraction of $\geq \varphi$ red children. By the inductive hypothesis the subspaces corresponding to them are $\beta$-approximately multilinear. There must, then, be a direction such that a fraction of $\geq \varphi$ of hyperplanes of $U_v$ in that direction is $\beta$-approximately multilinear.

Since $v$ is red, we also know that the proportion of wrong lines in $U_v$ is $< \epsilon_0$. This implies that the proposition of wrong lines in $U_v$ in any direction is $< n\epsilon_0$. Therefore, by the Pasting lemma $U_v$ is $\epsilon^*$-approximately multilinear, where $\epsilon^* = 3n^2(n\epsilon_0 + \delta + 1/N)$. This concludes the proof of the lemma since the choice of parameters implies that $\epsilon^* \leq \beta$. $\blacksquare$

$A$ is $\epsilon^*$-approximately multilinear. By the Self-improvement lemma it follows that $A$ is $\epsilon'$-approximately multilinear, completing the proof of Theorem 4.4.13. $\blacksquare$

The proof of Theorem 4.4.1 is immediate. Let our probabilistic Turing machine perform the test, setting the parameters so that $\epsilon' \leq \alpha$. If $A$ is multilinear, integral valued, and takes absolute values that are not too large, then the machine clearly accepts. On the other hand, Proposition 4.4.7 guarantees that if the machine accepts, condition 4.5 (Proposition 4.4.7) can be inferred with high confidence. By Theorem 4.4.13, this implies that $A$ is $\alpha$-approximately multilinear. ∎

## 4.4.7 Extensions

Our test for multilinearity is of interest independent of our application of the test. In this section we discuss the test further and look at variations of the test to different problems.

Our test assumes that the function $A$ is integral valued. But we can not be sure, if the test accepts that there exists, with high probability, an *integral* valued multilinear polynomial that interpolates $A$. Look, for example, at $A'(x_1, x_2, \ldots, x_n) := \frac{1}{2} x_1 \cdot x_2 \cdots x_n$. This function is not integral, when all the $x_i$s are odd, but observe that there is only an exponentially small fraction of points for which $A'$ is not integral. Hence if $A$ was equal to $\lfloor A' \rfloor$ then $A$ is $\frac{1}{2^n}$-approximately multilinear and our test accepts, with high probability. Therefore our test does not test for integral approximate multilinearity. But, on the other hand, it gives the following guarantee:

**Theorem 4.4.15** *Let*
$$P := \prod_{q \ prime} p^{m_q - 1},$$
*where $m_q = \left\lceil \frac{2n}{(q-1)(1 - 2n/N)} \right\rceil$.*

*If $A$ is $\frac{1}{4}$-approximated by a multilinear function $g$ and $P \cdot g$ is not integral valued, then $M^A$ rejects with probability at least $1 - p$.*

**Proof:** The conclusion follows from the claim below that states that a multilinear $g$, such that $Pg$ is not integral, is non-integral on a large fraction ($> \frac{1}{2}$) of $\mathbf{I}^n$, and since $A$ $\frac{1}{4}$-approximates $g$, $A$ is non-integral on a set of measure at least $\frac{1}{4}$. This implies that $M^A$ rejects with probability at least $1 - \left(\frac{3}{4}\right)^{n \ln(2/p)/\delta} > 1 - p$, because $M^A$ chooses $n \ln(2/p)/\delta$ random, uniformly distributed points and if it finds a point where $A$ is non-integral it rejects.

**Claim 1** *Let g be a multilinear polynomial in n variables. Assume that for some $\gamma \in \mathbf{I}^n$, $g(\gamma) = \frac{a}{q^m}$, where a is an integer, q is a prime, $m \geq \frac{2n}{(q-1)(1-2n/N)}$ and $q \nmid a$. Then*

$$\mu(\{\alpha \in \mathbf{I}^n \mid g(\alpha) \text{ is integral}\}) \leq \frac{1}{2}. \qquad (4.6)$$

To prove the claim assume that for all $\alpha \in \mathbf{I}^n$, $g(\alpha) = \frac{b}{q^i}$, for some integers $b$ and $i$. This is no restriction since we can always find an integer $k$ such that $kg$ satisfies the assumption 4.6, which implies that $g$ satisfies the assumption 4.6. In other words we are in a situation where for all $\alpha \in \mathbf{I}^n$, $g(\alpha)$ is a rational number with the denominator a power of $q$. We can partition $\mathbf{I}^n$ according to the power of the denominator. In other words

$$S_0 := \{\alpha \in \mathbf{I}^n \mid g(\alpha) \text{ is integral}\}$$

$$S_i := \left\{\alpha \in \mathbf{I}^n \;\middle|\; g(\alpha) = \frac{b}{q^i} \text{ for some integer } b \text{ such that } q \nmid b\right\}.$$

We know that $S_m$ is non-empty. Furthermore, define $T_i := \bigcup_{k=0}^{i} S_k$.

Given $\beta \in \mathbf{I}^n$ let $s(\beta)$ be the integer such that

$$\beta \in S_{s(\beta)}.$$

Look at $\mathbf{I}^n$ as a graph on $N^n$ nodes, where the graph is a tree rooted at $\gamma$. For each $\beta \in \mathbf{I}^n$ we define a path from $\gamma$ to $\beta$ by

$$\gamma = \beta^{(0)} \rightarrow \beta^{(1)} \rightarrow \cdots \rightarrow \beta^{(n)} = \beta,$$

where $\beta^{(k)} = (\beta_1, \beta_2, \ldots, \beta_k, \gamma_{k+1}, \ldots, \gamma_n)$ for $k \in \{0, 1, \ldots, n\}$.

Observe that $s(\beta^{(0)}) = m$ whereas for $\beta$ to be integral $s(\beta^{(n)}) = 0$. Define a measure $\delta_k$ of how much closer $\beta^{(k)}$ is to being integral compared to $\beta^{(k-1)}$:

$$\delta_k(\beta) := \max\left\{s(\beta^{(k-1)}) - s(\beta^{(k)}), 0\right\}.$$

Hence if $\beta \in S_0$ then

$$\sum_{k=1}^{n} \delta_k(\beta) \geq m.$$

Therefore if $g$ is integral on at least half of $\mathbf{I}^n$ then

$$\sum_{\beta} \sum_{k=1}^{n} \delta_k(\beta) \geq N^n m / 2. \tag{4.7}$$

In the rest of the proof we give a lower bound on the same summation and get a bound for $m$.

For $\alpha = \beta^{(k-1)} \in S_i$ we look at the line $\ell$ in direction $k$ through $\alpha$. Since $g$ is multilinear

$$g|_\ell(x) = \frac{c_1}{q^j}(x_i - \beta_i) + \frac{c_1}{q^i},$$

where $c_1$, $c_2$ and $j$ are integers. This linear function has a simple form because of our assumption about $g$.

If $j < i$ then for all $\alpha' \in \ell$, $\alpha' \in S_i$, since $f(\alpha')$ is always of the form $\frac{b}{q^i}$, where $b$ is an integer that is relatively prime to $q$.

If $j \geq i$ then $\alpha' \in T_d$ for $d < i$ if and only if

$$c_1 \alpha'_k + q^{j-i} c_2 \equiv 0 \pmod{q^{i-d}}.$$

Hence,

$$|\{\alpha' \in \ell \mid \alpha' \in T_d\}| \leq 1 + \lfloor (N-1)q^{-(i-d)} \rfloor.$$

We need to bound the sum of $\delta_k$ over $\alpha' \in \ell$:

$$\begin{aligned}
\sum_{\alpha' \in \ell} \delta_k(\alpha') &= \sum_{\alpha' \in \ell} \max\{s(\alpha) - s(\alpha'), 0\} \\
&= \sum_{\alpha' \in \ell} \sum_{d=1}^{i} [\alpha' \in T_{i-d}] \\
&= \sum_{d=1}^{i} \sum_{\alpha' \in \ell} [\alpha' \in T_{i-d}] \\
&\leq \sum_{d=1}^{i} 1 + \lfloor (N-1)q^{-d} \rfloor \\
&< i + (N-1)/(q-1).
\end{aligned}$$

This give us the lower bound on the sum in Equation 4.7.

$$\sum_{\beta} \sum_{k=1}^{n} \delta_k(\beta) = \sum_{k=1}^{n} \sum_{\beta} \delta_k(\beta)$$

$$= \sum_{k=1}^{n} \sum_{\beta_1,\ldots,\beta_{k-1}} \sum_{\alpha' \in \ell} N^{n-k} \delta_k(\alpha')$$

$$< N^{n-k} \sum_{k=1}^{n} \sum_{\beta_1,\ldots,\beta_{k-1}} (m + N/(q-1))$$

$$= n N^{n-1}(m + N/(q-1)).$$

Therefore we get that

$$m < \frac{2n}{(q-1)(1-2n/N)},$$

which contradicts our assumption about $m$. ∎

It is easy to see that our test also works when $A$ is defined over finite fields.

**Theorem 4.4.16** *Given a finite field $\mathbf{F}_q$, let $\alpha$ and $p$ be positive reals and let $n$ be an integer such that $\frac{40n^2}{\alpha} < q$. Let $A(x_1,\ldots,x_n)$ be an arbitrary function from $\mathbf{F}_q^n$ to $\mathbf{F}_q$. Then there exists a probabilistic Turing machine $M$, which works in time polynomial in $n, \log q, \log \frac{1}{p}$ and $\frac{1}{\alpha}$, such that, given access to $A$ as an oracle,*

- *If $A$ is multilinear then $M^A$ always accepts.*

- *If $A$ is not $\alpha$-approximately multilinear then $M^A$ rejects with probability at least $1 - p$.*

**Proof:** The proof of Theorem 4.4.1 goes through with only minor changes. $\mathbf{F}_q$ takes the role of $\mathbf{I}$ and $q$ the role of both $K$ and $N$.

A variation of our test also works if we replace multilinearity by the condition that the polynomial be of low degree. Let $k_1,\ldots,k_n$ be positive integers. Assume we wish to test if the function $A : \mathbf{I}^n \to \mathbf{Q}$ is a polynomial having degree $\leq k_i$ in variable $x_i$, for every $i$. Lemma 4.3.3 extends to this situation, with only trivial changes in the proof.

**Theorem 4.4.17** *Let $\alpha$ and $p$ be positive reals and let $n, k_1, k_2, \ldots, k_n, K$ and $N$ be integers such that $\frac{20n^2(k+1)^2}{\alpha} < N$, where $k = \max_i k_i$. Let $\mathbf{I}$ denote the set of integers $\{0, \ldots, N-1\}$. Let $A(x_1, \ldots, x_n)$ be an arbitrary function from $\mathbf{I}^n$ to $\mathbf{Q}$. Then there exists a probabilistic Turing machine $M$, which works in time polynomial in $n, k, \log N, \log \frac{1}{p}, \frac{1}{\alpha}$ and $\log K$, such that, given access to $A$ as an oracle,*

- *If $A$ is polynomial such that the degree of $x_i$ is at most $k_i$ for all $i$ and $A$ is integral valued and does not take absolute values greater than $K$, then $M^A$ always accepts.*

- *If there does not exist a polynomial $g$ such that the degree of $x_i$ is at most $k_i$ for all $i$ and $g$ $\alpha$-approximates $A$, then $M^A$ rejects with probability at least $1 - p$.*

**Proof:** We modify our test such that for each direction $i$ we test that $A$ restricted to lines in the $i$th direction can be interpolated by polynomials of degree at most $k_i$. That this test satisfies the conclusions follows since Theorem 4.4.13 extends to this situation, with the following change of the parameters. We should set $\epsilon' = 3n^2(\epsilon + \delta + k/N)$ where $k = \max_i k_i$; furthermore $\beta = \frac{1}{5(k+1)}$, $N \geq 20(k+1)^2 n^2$, $\delta \leq \frac{1}{200n^2(k+1)}$ and $\epsilon \leq \frac{1}{400n^3(k+1)}$. With the obvious changes, the same proof applies. ∎

There is a similar theorem for the case in which $A$ is defined over a finite field instead of over the integers.

We can also extend our test to cover the case where we have to test if the function $A$ is a polynomial of degree at most $k$. Here we state the theorem for functions over finite fields. The theorem holds for the integral case too, but the proof is more tedious.

**Theorem 4.4.18** *Given a finite field $\mathbf{F}_q$. Let $\alpha$ and $p$ be positive reals and let $n$ and $k$ be integers such that $\frac{20n^2(k+1)^2}{\alpha} < q$. Let $A(x_1, \ldots, x_n)$ be an arbitrary function from $\mathbf{F}_q^n$ to $\mathbf{F}_q$. Then there exists a probabilistic Turing machine $M$ which works in time polynomial in $n, k, \log \frac{1}{p}, \frac{1}{\alpha}$ and $\log q$, such that, given access to $A$ as an oracle,*

- *If A is a polynomial of degree at most k then $M^A$ always accepts.*

- *If there does not exist a polynomial g of degree at most k such that g α-approximates A then $M^A$ rejects with probability at least $1 - p$.*

**Proof:** From the finite field version of Theorem 4.4.17 we can test whether there exists a polynomial $g$ such that the degree of every variable is at most $k$ and $g$ α-approximates $A$ for $\alpha < \frac{1}{3nk+3}$. We run the test such that we know $g$ exists with probability at least $1 - \frac{p}{2}$. We are left with the problem of testing whether $g$ has degree at most $k$. We can write $g$ as

$$g(x_1, x_2, \ldots, x_n) = \sum_{\substack{d_1, d_2, \ldots, d_n \\ 0 \le d_i \le k}} a_{d_1, d_2, \ldots, d_n} x_1^{d_1} x_2^{d_2} \cdots x_n^{d_n}.$$

Hence $g$ has degree $m \le k$ if and only if for all $d_1, d_2, \ldots, d_n$ such that $d_1 + d_2 + \cdots + d_n > k$, $a_{d_1, d_2, \ldots, d_n} = 0$.

For $\beta = (\beta_1, \beta_2, \ldots, \beta_n)$ define a polynomial $h_\beta(x)$ by

$$h_\beta(x) := g(\beta_1 x, \beta_2 x, \ldots, \beta_n x)$$

and

$$h_\beta(x) = \left( \sum_{\substack{d_1, d_2, \ldots, d_n \\ \sum_i d_i = m}} a_{d_1, d_2, \ldots, d_n} \beta_1^{d_1} \beta_2^{d_2} \cdots \beta_n^{d_n} \right) x^m + h'(x),$$

where $h'(x)$ is a polynomial of degree at most $m - 1$. So the degree of $h_\beta(x)$ is $m$ unless $\beta$ is a root in the nonzero multivariate polynomial

$$\sum_{\substack{d_1, d_2, \ldots, d_n \\ \sum_i d_i = m}} a_{d_1, d_2, \ldots, d_n} x_1^{d_1} x_2^{d_2} \cdots x_n^{d_n}.$$

Since any nonzero multivariate polynomial has at most $q^{n-1} \deg(g) \le q^{n-1} nk$ [65] roots, if we choose $\beta$ at random and uniformly from $\mathbf{F}_q^n$, then the degree of $h_\beta(x)$ is $\deg(g)$, with probability at least $1 - \frac{nk}{q}$. To find the degree of $h_\beta(x)$ we evaluate $h_\beta(x)$ for $x = 1, 2, \ldots, nk+1$ and interpolate $h_\beta(x)$ given the values.

The test evaluates $A$ at the points $(\beta_1 i, \beta_2 i, \ldots, \beta_n i)$ for $i = 1, 2, \ldots, nk+1$, interpolates a polynomial $f$ such that $f(i) = A(\beta_1 i, \beta_2 i, \ldots, \beta_n i)$ for all $i$ and checks that the degree of $f$ is at most $k$. If $g$ and $A$ agree on these points then we correctly calculate the degree of $h_\beta(x)$. The probability that $A$ and $g$ agree on these $nk + 1$ points is at least $1 - \frac{nk+1}{\alpha}$, since each point is a random point uniformly distributed in $\mathbf{F}_q^n$ and because $A$ $\frac{1}{\alpha}$-approximate $g$.

If $\deg(g) > k$, then with probability at least $1 - \frac{nk}{q} - \frac{nk+1}{\alpha} > \frac{1}{2}$ we discover so. Therefore if we choose $\log_2(\frac{2}{p})$ independent $\beta$s, with probability at least $1 - \frac{p}{2}$ we discover that $\deg(g) > k$. ∎

# 5

## Interaction versus Alternation

In this chapter we look at the relationship between alternating Turing machines and interactive proof systems.

There are some obvious similarities between alternating Turing machines and interactive proof systems. In fact, Goldwasser and Sipser [41] proved the equivalence of interactive proof systems and polynomial time Turing machines alternating between nondeterministic and probabilistic moves. However, until the results described in Chapter 3 appeared, it was generally believed that alternating Turning machines had far more power than interactive proof systems. The results in Chapter 3 show that the set of languages accepted by an interactive proof system equals the class of languages accepted in deterministic polynomial space. Since Chandra, Kozen and Stockmeyer [20] have shown *PSPACE* to be equivalent to the languages accepted by a polynomial time alternating Turing machine, in this case, alternating Turing machine and interactive proof systems have identical power.

We generalize the work in Chapter 3 to exhibit a broader equivalence between alternating Turing machines and interactive proof systems. We look at time-space complexity, first studied for alternating Turing machines by Ruzzo [63] and for interactive proof systems by Condon [21].

We study only interactive proof systems with public coins. Goldwasser and Sipser [41] prove that the class of languages accepted by interactive proofs with a polynomial time verifier does not depend on whether the verifier uses public or private coins. However, a difference between private and public coins does hold for time- and space-bounded verifiers as we saw in Section 2.3.5.

We exhibit a general relationship between time- and space-bounded alternating Turing machines and time- and space-bounded public-coin verifiers. We exhibit that all languages accepted by an interactive proof system with a $t(n)$-time, $s(n)$-space bounded verifier can also be accepted by an alternating Turing machine using $t(n)\log t(n)$-time and $s(n)$-space. Conversely, we show that an interactive proof system can simulate any $t(n)$-time, $s(n)$-space alternating Turing machine using $\text{poly}(n)+\text{poly}(t(n))$-time and $\text{poly}(s(n))$-space bounded verifier.

We use this close relationship between alternating Turing machines and interactive proof systems to exhibit a public-coin interactive proof system for all languages in *NC* with a polynomial time log-space verifier. Furthermore we exhibit a proof system for all languages in P with a polynomial time verifier that uses less than log-squared space. The previous best result known, by Fortnow and Sipser [34], proves that *LOGCFL* has a public-coin interactive proof system with a polynomial time log-space verifier. *LOGCFL* consists of all languages log-space reducible to context-free languages and is known to lie in $NC^2$ [71, 72, 63]. Furthermore, we use this relationship to obtain a hierarchy for public-coin interactive proof systems.

We also use these theorems to get strong relationships between interactive proof systems and deterministic computation similar to the relationships between alternating Turing machine and deterministic computation found in Section 2.3.4 due to Chandra, Kozen and Stockmeyer [20].

# 5.1  Restricted Alternating Turing Machines

In this chapter we use the "random access input" model for an alternating Turing machine similar to the one described by Ruzzo [63]. This allows us to study alternating Turing machines that use sublinear time. In our model, the alternating machine $M$ has two special states, $q_0$ and $q_1$. When $M$ enters the state $q_j$ with a value $i$ written in binary on its first work tape, then $M$ accepts if the $i$th bit of the input is $j$ and rejects otherwise. Note that we can simulate arbitrary access to the input by guessing the value of the input and universally verifying that value. We, additionally, assume that both the verifier and the alternating Turing machine have some constant number $k$ of read-write tapes, each with its own head.

To arithmetize alternating computation efficiently, we introduce a special type of alternating Turing machine. It is a restricted model, but we use the rest of this section to show that it is not a restriction in computational power.

First we restrict the number of tapes to one and we prove that this does not decrease the computational power.

Paul, Prauss and Reischuk in [60] proved such a theorem for $ATIME(t)$.

**Theorem 5.1.1 (Paul, Prauss and Reischuk [60].)** *Let $L$ be a language in $ATIME(t(n))$. There exists a 1-tape alternating Turing machine $M$ such that $M$ works in time $O(t(n))$ and $L(M) = L$.*

We need to extend their result to $ATISP(t, s)$. Recall that $ATISP(t, s)$ is the class of languages that is recognized by alternating Turing machines in time $t(n)$ and space $s(n)$. From their proof it is clear that their simulation uses $O(t(n))$ space. We present a simulation of a $k$-tape alternating Turing machine that works in time $t(n)$ and space $s(n)$ by a 1-tape alternating Turing machine that works in time $O(t(n))$ and space $O(s(n))$. In our proof we use both their result and the ideas of their proof.

**Theorem 5.1.2** *Let $L \in ATISP(t(n), s(n))$. There exists a 1-tape alternating Turing machine $M$ that works in time $O(t(n))$ and space $O(s(n))$ and $L(M) = L$.*

**Proof:** Let $N$ be a $k$-tape alternating Turing machine that recognizes $L$ such that $N$ works in time $O(t(n))$ and space $O(s(n))$. We construct $M$ such that it simulates $N$ in *phases* corresponding to time blocks of size $s(n)$. At the beginning of each phase $M$ has the contents of $N$'s tapes stored on a track on its work tape. $M$ starts each phase by guessing the next $s(n)$ *displays* of $N$ on a second track of its work tape. The display of $N$ at time $i$ consists of the state and the content of each of the $k$ cells that $N$ scans at time $i$. $M$ is in a universal (respectively existential) state if $N$ is in a universal (respectively existential) state. On a third track it guesses existentially the content of $N$'s tapes after the phase. Note that if the displays correspond to a valid computation path of $N$ then there exist such tape contents, which is unique.

Thereafter, $M$ checks the validity of its guesses. Observe that checking the validity of the guesses can easily be done on a $k+1$ tape deterministic Turing machine $M_1$ in time $O(s(n))$. Because of Theorem 5.1.1 $M$ can simulate $M_1$ using one tape in time $O(s(n))$. If the guesses correspond to a computation path in $N$ then $M$ starts a new phase. If not, it has to figure out which type of state it was in when it made the first wrong guess. Note that $M$ has to reject (respectively accept) if it guessed wrong in an existential (respectively universal) state. But a $k+1$ tape deterministic Turing machine $M_2$ can in time $O(s(n))$ easily find the type of the state, where the first wrong guess was made. $M$ can simulate $M_2$ in time $O(s(n))$ using Theorem 5.1.1. Furthermore, if at some point $N$ accepts or rejects, $M$ does the same. It is easy to see that $L(M) = L(N)$ and that $M$ works in time $O(t(n))$ and space $O(s(n))$. ∎

Next we restrict our model even further. The alternating Turing machine first makes an existential move consisting of two possible moves, followed by a universal move of two possible moves and so on. The computation tree becomes a binary tree with alternating levels of AND gates and OR gates. Furthermore, we assume that all computation paths have even length. It is easy to see that given an arbitrary alternating Turing machine $M$ we can construct an alternating Turing machine $N$ that has a computation tree as described above and that $N$ works in time and space proportional to the time and space used by $M$.

The machines we consider are *restricted* Alternating Turing machines $M$ such that

- $M$ has only one tape, which at the start of the computation contains the binary representation of $n = |x|$.

- $M$'s computation tree is a binary tree with alternating levels of AND gates and OR gates and all computation paths have even length.

- On each computation path $M$ only reads one input bit and does that at the end. A computation path accepts or rejects depending on the $i$th bit of the input, if the binary representation of $i$ is stored in the first

$\lceil \log n \rceil$ cells on the work tape at the end of the computation, otherwise it rejects.

## 5.2   Arithmetization of Alternating Computation

Let $L \in ATISP(t, s)$ and let $M$ be a restricted alternating Turing machine such that $L(M) = L$. We assume, without loss of generality, that $M$ works in time $t(n)$ and uses $s(n)$ space. Let $\Delta$ be the work alphabet and let $Q$ be the set of $M$'s states. Given an input string $x$, we define an arithmetic expression $E_x$ such that the value of $E_x$ determines if $x \in L$. We construct $E_x$ in this section and in the next section we show how an interactive proof system can verify the value of $E_x$. Thereafter we follow the general outline in the previous two chapters.

Let $\varphi_i(I, x)$ be the predicate that is true if and only if, from the configuration described by the ID $I$, in at most $2i$ steps $M$ accepts $x$. We assume that $I$ describes $M$ to be in an existential state. Let $M(I)$ denote the machine $M$ in the configuration described by $I$. Because of the simple structure of the computation of the restricted machine $M$, it is clear for $i > 0$ that $M(I)$ accepts $x$ in at most $2i$ steps if

- $M(I)$ makes an existential guess $a$ followed by a universal guess $b$ and ends up having an ID $I'$ such that $M(I')$ accepts $x$ in at most $2(i - 1)$ steps.

We make the assumption that when $M$ enters a final configuration it stays in this configuration indefinitely. For $i = 0$, $M(I)$ accepts $x$ if and only if $I$ describes an accepting configuration. This gives an inductive definition of $\varphi_i(I, x)$:

$$\varphi_i(I, x) = \begin{cases} g_x(I) & \text{if } i = 0 \\ \exists a \forall b \exists I' : f(I, I', a, b) \wedge \varphi_{i-1}(I', x) & \text{otherwise,} \end{cases}$$

where $f(I, I', a, b)$ is the predicate stating that $M(I)$, on input $x$, on an existential guess $a$ and a universal guess $b$, gets into the configuration described by $I'$. The predicate $g_x(I)$ states that $M$ accepts input $x$ if in configuration $I$.

If $N = \lceil t(n)/2 \rceil$ and $I_0$ is the ID describing the start configuration then clearly

$$\varphi_N(I_0, x) = 1 \iff M \text{ accepts } x.$$

We extend the ideas in Chapter 3 to arithmetize $\varphi_N(I_0, x)$. Given a Boolean $\varphi$ expression we construct an arithmetic expression $A_\varphi$. The following table inductively defines $A_\varphi$:

| $\varphi$ | $A_\varphi$ |
|:---:|:---:|
| $0$ | $0$ |
| $1$ | $1$ |
| $x_i$ | $x_i$ |
| $\neg\varphi'$ | $1 - A_{\varphi'}$ |
| $\varphi' \wedge \varphi''$ | $A_{\varphi'} A_{\varphi''}$ |
| $\varphi' \vee \varphi''$ | $1 - (1 - A_{\varphi'})(1 - A_{\varphi''})$ |
| $\forall x : \varphi'(x)$ | $\prod_{x=0}^{1} A_{\varphi'}(x)$ |
| $\exists x : \varphi'(x)$ | $\coprod_{x=0}^{1} A_{\varphi'}(x)$ |

where $\coprod_{x=0}^{1} A_{\varphi'}(x)$ is short-hand for $1 - \prod_{x=0}^{1}(1 - A_{\varphi'}(x))$.

In what follows we use these rules except for a slight change in the case where $\varphi = \exists x : \varphi'(x)$. Sometimes we let $A_\varphi = \sum_{x=0}^{1} A_{\varphi'}(x)$. This is in the case where we know that $\varphi'(x)$ is true, for, at most, one value of $x$. The advantage is that the degree of the polynomial $A_\varphi$ is, at most, the degree of $A_{\varphi'}$ whereas in the original case the degree doubles. It is crucial to keep the degree low as we will see later.

Using these rule we get that

$$A_\varphi = \begin{cases} 1 & \text{if } \varphi \text{ is true} \\ 0 & \text{otherwise.} \end{cases}$$

This can easily be proven by induction.

We now need an encoding of IDs to arithmetize $f(I, I', a, b)$ and $g_x(I)$.

Our encoding of IDs is a compact version of a binary encoding. The binary encoding encodes $I$ as a tuple $(q, c_1, c_2, \ldots, c_{s(n)}, h_1, h_2, \ldots, h_{s(n)})$, where

- $q = (q_1, q_2, \ldots, q_{k'})$ is a binary encoding of the state of $M$, where $k' = \lceil \log_2(|Q|) \rceil$.

- $c_i = (c_{i1}, c_{i2}, \ldots, c_{ik})$ is a binary encoding of the content of the $i$th cell, where $k = \lceil \log_2(|\Delta|) \rceil$.

- $h_i$ is 1 if and only if the head of $M$ is scanning the $i$th cell.

We describe how to arithmetize $\varphi_N(I_0, x)$ using the binary encoding and later we extend the arithmetization to a more compact encoding.

To arithmetize the computation of $M$ we first assume that the head is scanning the $i$th cell. Let $I = (q, c_1, \ldots, h_{s(n)})$ and $I' = (q', c_1', \ldots, h_{s(n)}')$. Given the head position, it is locally decidable if $I'$ follows $I$ on the guesses $a$ and $b$. Define

$$\varphi_{mi} := [c_m' \text{ is correct given that the head scans cell } i.]$$

$$\eta_{mi} := [h_m' \text{ is correct given that the head scans cell } i.]$$

$$\psi_i := [q' \text{ is correct given that the head scans cell } i.]$$

Next we note which variables the above predicates depend on:

- $\varphi_{mi}$ depends only on $c_m$ and $c_m'$ if $m \notin \{i-1, i, i+1\}$, since in two steps $M$ can change only the content of cells $i-1, i, i+1$. For $m \in \{i-1, i, i+1\}$, $\varphi_{mi}$ depends on $q, c_{i-1}, c_i, c_{i+1}, a, b$ and $c_m'$.

- $\eta_{mi}$ depends only on $h_m'$ if $m \notin \{i-2, i-1, i, i+1, i+2\}$, since in two steps $M$ can move its head only two positions. For $m \in \{i-2, i-1, i, i+1, i+2\}$, $\eta_{mi}$ depends on $q, c_{i-1}, c_i, c_{i+1}, a, b$ and $h_m'$.

- $\psi_i$ depends on the variables $q, c_{i-1}, c_i, c_{i+1}, a, b$ and $q'$.

Let $f_i$ be the predicate that is true if and only if $I'$ follows $I$ on the guesses $a$ and $b$, given that the head is scanning the $i$th cell. From the above definitions we can write down a simple Boolean formula for $f_i$. We get that $f_i$ is true if and only if

$$\psi_i \wedge \bigwedge_{m=1}^{s(n)} \left(\varphi_{mi} \wedge \eta_{mi}\right)$$

is true.

But since $h_i$ describes if the head is at cell $i$ in $I$, $f$ can be written as

$$\bigvee_{i=1}^{s(n)} \left(h_i \wedge f_i\right).$$

Given this description of $f$, it is straightforward to arithmetize $f$. By Proposition 2.1.1 any predicate $\gamma$ can be interpolated by a multilinear polynomial $P_\gamma$. If $\gamma$ only depends on a fixed number of variables then $P_\gamma$ can be computed by a formula $F_\gamma$ of fixed size. We obtain formulas $F_{\varphi_{mi}}, F_{\eta_{mi}}$ and $F_{\psi_i}$ that compute multilinear polynomials that interpolate $\varphi_{mi}, \eta_{mi}$ and $\psi_i$ respectively.

This gives arithmetic formulas that compute polynomials that interpolate $f_i$ and $f$:

$$F_{f_i} := F_{\psi_i} \prod_{m=1}^{s(n)} F_{\varphi_{mi}} F_{\eta_{mi}}$$

$$F_f := \sum_{i:=1}^{s(n)} h_i F_{f_i}.$$

**Lemma 5.2.1** *For any $I, I' \in \{0,1\}^{(k+1)s(n)+k'}$, where $I$ is an encoding of a valid ID, and for any $a, b \in \{0,1\}$,*

$$F_f(I, I', a, b) := \begin{cases} 1 & \text{if } f(I, I', a, b) \text{ is true,} \\ 0 & \text{otherwise.} \end{cases}$$

*Furthermore the degree of any variable in $F_f$ is at most 9, $size(F_f) = O(s^2(n))$ and $depth(F_f) = O(\log s(n))$.*

**Proof:** That $F_f$ interpolates $f$ is clear when we observe that for at most one $i$, $h_i \wedge f_i$ is true. The reason is that for, at most, one $i$, $h_i$ is true, since $I$ is an encoding of a valid ID. Thus, at most, one summands of $F_f$ is one.

To find the degree of, say, $q_1$, note that in $f_i$, it is only $\psi_i, \varphi_{i-1,i}, \varphi_{i,i}, \varphi_{i+1,i}, \eta_{i-2,i}, \eta_{i-1,i}, \eta_{i,i}, \eta_{i+1,i}$ and $\eta_{i+2,i}$ that depend on $q_1$. Note that $q_1$ is only a variable in the corresponding 9 formulas, each of which is multilinear. The degree of $q_1$ is at most 9 in $F_{f_i}$ and therefore in $F_f$. The following table gives a complete description of the bounds on the degrees for every variable.

|         | $q_j$ | $q'_j$ | $c_{mj}$ | $c'_{mj}$ | $h_m$ | $h'_m$ | $a$ | $b$ |
|---------|-------|--------|----------|-----------|-------|--------|-----|-----|
| $F_{f_i}$ | 9     | 1      | 9        | 1         | 0     | 1      | 9   | 9   |
| $F_f$   | 9     | 1      | 9        | 1         | 1     | 1      | 9   | 9   |

This completes the proof. ∎

To arithmetize $g_x$ observe that, because $M$ is restricted, $g_x$ depends only on $c_1, c_2, \ldots, c_{\lceil \log n \rceil}, q$ and $x$. In order for $I$ to be an accepting ID $c_1, c_2, \ldots, c_{\lceil \log n \rceil}$ should be the binary representation of some number $j \in \{1, 2, \ldots, n\}$ and furthermore, the input bit $x_j$ should be 0 or 1 depending on $q$. So for each $j$ and $l \in \{1, 2, \ldots, \lceil \log n \rceil\}$ we define $\sigma_{jl}(c_l)$ to be true if and only if $c_l$ encodes the $l$th bit in the binary representation of $j$. Furthermore, we define $\rho_j(q, x_j)$ to be true if and only if in state $q$, $M$ immediately accepts if the $j$th input is $x_j$.

$$g_x = \bigvee_{j=1}^{n} \left( \bigwedge_{l=1}^{\lceil \log n \rceil} \sigma_{jl}(c_l) \right) \wedge \rho_j(q, x_j).$$

This gives us the following arithmetic formula,

$$F_{g_x} := \sum_{j=1}^{n} \left( \prod_{l=1}^{\lceil \log n \rceil} F_{\sigma_{jl}}(c_l) \right) F_{\rho_j}(q, x_j),$$

where $F_{\sigma_{jl}}$ and $F_{\rho_j}$ are formulas that compute multilinear polynomials that interpolate $\sigma_{jl}$ and $\rho_j$ respectively.

**Lemma 5.2.2** *For any $I \in \{0,1\}^{(k+1)s(n)+k'}$, where $I$ is an encoding of a valid ID,*

$$F_{g_x}(I) = \begin{cases} 1 & \text{if } g_x(I) \text{ is true,} \\ 0 & \text{otherwise.} \end{cases}$$

*Moreover the degree of any variable in $F_{g_x}$ is at most 1, $\text{size}(F_{g_x}) = O(n \log n)$ and $\text{depth}(F_{g_x}) = O(\log n)$.*

**Proof:** Clear, since there exists at most one $j$ such that

$$\left( \bigwedge_{l=1}^{\lceil \log n \rceil} \sigma_{jl}(c_l) \right) \wedge \rho_j(q, x_j)$$

is true. $\blacksquare$

We improve the space efficiency of the encoding of IDs by using a more compact encoding. Choose an $\epsilon > 0$ and encode $m = \lfloor \frac{\epsilon}{2} \log s(n) \rfloor$ binary symbols $\alpha_0, \alpha_1, \ldots, \alpha_{m-1}$ into a new symbol $\alpha \in \{0, 1, \ldots, 2^m - 1\}$, just by letting $\alpha = \sum_{i=0}^{m-1} \alpha_i 2^i$. Let $X = \{0, 1, \ldots, 2^m - 1\}$. To decode the new symbol define $D_i : X \to \{0, 1\}$ that maps an $m$-bit number $\alpha$ into the $i$th bit of $\alpha$. The decoding is obtained, using Lagrange interpolation, by the following formula:

$$F_{D_i}(x) = \sum_{u \in X} D_i(u) \prod_{v \neq u} \frac{(x - v)}{(u - v)}.$$

The degree of the polynomial computed by $F_{D_i}$ is $2^m - 1$, and $\text{size}(F_{D_i}) = O(2^{2m})$ and $\text{depth}(F_{D_i}) = O(m)$. Define

$$F(y_1, y_2, \ldots, y_{n'}, y_1', y_2', \ldots, y_{n'}', a, b) :=$$
$$F_f(F_{D_1}(y_1), F_{D_2}(y_1), \ldots, F_{D_m}(y_1), F_{D_1}(y_2), \ldots, F_{D_m}(y_{n'}'), a, b)$$

where $n' = \lceil \frac{(k+1)s(n)+k'}{m} \rceil$. In other words we modify the formula $F_f$. For each leaf $l$ containing a variable $z$ that encodes part of an ID, we are replacing $l$ with the subformula that decodes $z$ from the appropriate variable of $F$. For example, if $z$ is $z_1$, the first binary variable that encodes $I$, then the subformula is $F_{D_1}(y_1)$, since $y_1$ encodes $z_1$.

**Lemma 5.2.3** *For any $I, I' \in X^{n'}$, where $I$ is an encoding of a valid ID, and for any $a, b \in \{0,1\}$,*

$$F(I, I', a, b) = \begin{cases} 1 & \text{if } f(I, I', a, b) \text{ is true,} \\ 0 & \text{otherwise.} \end{cases}$$

*Moreover the degree of any variable in $F$ is at most $9(2^m - 1) < 9s^{\epsilon/2}(n)$, $\text{size}(F) = O(2^{2m} s^2(n)) = O(s^{2+\epsilon}(n))$ and $\text{depth}(F) = O(\log s(n))$.*

**Proof:**  Follows from Lemma 5.2.1.  ∎

We define $G_x$ in a similar way from the definition of $F_{g_x}$.

$$G_x(y_1, ..., y_{n'}) := F_{g_x}(F_{D_1}(y_1), F_{D_2}(y_1), ..., F_{D_m}(y_1), F_{D_1}(y_2), ..., F_{D_m}(y_{n'})).$$

**Lemma 5.2.4** *For any $I \in X^{n'}$, where $I$ is an encoding of a valid ID,*

$$G_x(I) = \begin{cases} 1 & \text{if } g_x(I) \text{ is true,} \\ 0 & \text{otherwise.} \end{cases}$$

*Furthermore the degree of any variable in $G_x$ is at most $(2^m - 1) < s^{\epsilon/2}(n)$, $\text{size}(G_x) = O(2^{2m} n \log n) = O(s^{\epsilon}(n) n \log n)$ and $\text{depth}(G_x) = O(\log s(n) + \log n)$.*

**Proof:**  Clear, from Lemma 5.2.2.  ∎

With the arithmetization of $f$ and $g_x$, we get an arithmetization of $\varphi_i(I, x)$ by inductively defining

$$A_i(I, x) = \begin{cases} G_x(I) & \text{if } i = 0, \\ \coprod_{a=0}^{1} \prod_{b=0}^{1} \sum_{I' \in X^{n'}} F(I, I', a, b) \cdot A_{i-1}(I', x) & \text{otherwise,} \end{cases}$$

where $X = \{0, 1, \ldots, 2^m - 1\}$. Remember that $\coprod_{x=0}^{1} q(x) = 1 - \prod_{x=0}^{1}(1 - q(x))$.

**Lemma 5.2.5** *If $I_0 = (y_1, y_2, \ldots, y_{n'}) \in X^{n'}$ is the encoding of $M$'s starting configuration then*

$$A_N(I_0, x) = \begin{cases} 1 & \text{if } x \in L, \\ 0 & \text{otherwise.} \end{cases}$$

**Proof:** We prove by induction that for all $i$ and for all encodings $I$ of valid IDs,

$$A_i(I, x) = \begin{cases} 1 & \text{if } \varphi_i(I, x) \text{ is true,} \\ 0 & \text{otherwise.} \end{cases}$$

For $i = 0$ this follows from Lemma 5.2.4.

For $i > 0$, observe that there is one $I'$ such that $f(I, I', a, b)$ is true. Observe that both $I$ and $I'$ are encodings of valid IDs. It follows from Lemma 5.2.3 and the inductive hypothesis that

$$A_i(I, x) = \begin{cases} 1 & \text{if } \varphi_i(I, x) \text{ is true,} \\ 0 & \text{otherwise.} \end{cases} \quad \blacksquare$$

## 5.3 Interactive Proof Systems for $ATISP(t, s)$

Using the ideas in Chapter 3 it is straightforward to construct an interactive proof system that verifies that the value of $A_N(I_0, x)$ is 1. The verifier is working in the finite field of $p$ elements, where $p$ is a prime. We choose $p = O(s^{1.5}(n)t(n))$. The verifier eliminates the arithmetic operators using the technique from Chapter 3. See Figure 5.1 for all the details of the protocol.

We can then prove that protocol 6 recognizes $L$.

**Lemma 5.3.1** *Protocol 6 satisfies the following statements:*

1. *If $x \in L$ then there exists a prover such that the verifier always accepts.*

2. *If $x \notin L$ then for all provers, the verifier accepts with probability at most $\frac{1}{3}$.*

3. *The verifier works in time*

$$O((t(n)s^{2+\epsilon}(n) + s^\epsilon(n)n \log n) \log^2 t(n))$$

   *and uses space*

$$O\left(\frac{s(n) \log p}{\log s(n)}\right) = O\left(\frac{s(n) \log t(n)}{\log s(n)}\right).$$

**Protocol 6**
  **V:** $m \leftarrow \lfloor \frac{\epsilon}{2} \log s(n) \rfloor$, $N \leftarrow \lceil t(n)/2 \rceil$.
  **V:** $n' \leftarrow \lceil \frac{(k+1)s(n)+k'}{m} \rceil$, $d \leftarrow \lfloor 9s^{\epsilon/2}(n) \rfloor$.
**P→V:** $p \in [15dN(n'+2), 30dN(n'+2)]$ and a proof that $p$ is a prime.
  **V:** Lets $\alpha_1, \alpha_2, \ldots, \alpha_{n'}$ be the encoding of the start configuration.
  **V:** $\beta \leftarrow 1$.
    **for** $i = 1$ **to** $N$ **do**
        $\{$Eliminate $\coprod_a\}$
            **P→V:** a polynomial $q$ over $\mathbf{F}_p$ of degree at most $2d$.
            **V:** if $\coprod_{a=0}^{1} q(a) \neq \beta$ **then Halt** and **Reject**.
            **V:** Chooses $\xi_1$ uniformly at random in $\mathbf{F}_p$.
            **V:** $\beta \leftarrow q(\xi_1)$
        $\{$Eliminate $\prod_b\}$
            **P→V:** a polynomial $q$ over $\mathbf{F}_p$ of degree at most $d$.
            **V:** if $\prod_{b=0}^{1} q(b) \neq \beta$ **then Halt** and **Reject**.
            **V:** Chooses $\xi_2$ uniformly at random in $\mathbf{F}_p$.
            **V:** $\beta \leftarrow q(\xi_2)$
        **for** $j = 1$ **to** $n'$ **do**
            $\left\{$Eliminate $\sum_{y'_j \in X}\right\}$
                **P→V:** a polynomial $q$ over $\mathbf{F}_p$ of degree at most $5d$.
                **V:** if $\sum_{y'_j \in X} q(y'_j) \neq \beta$ **then Halt** and **Reject**.
                **V:** Chooses $\alpha'_j$ uniformly at random in $\mathbf{F}_p$.
                **V:** $\beta \leftarrow q(\alpha'_j)$
        **end**
        **V:** $\gamma \leftarrow F(\alpha_1, \ldots, \alpha_{n'}, \alpha_1, \ldots, \alpha_{n'}, \xi_1, \xi_2)$
        **V:** if $\gamma = 0$ and $\beta = 0$ **then Halt** and **Accept**.
        **V:** if $\gamma = 0$ and $\beta \neq 0$ **then Halt** and **Reject**.
        **V:** if $\gamma \neq 0$ **then**
            $\beta \leftarrow \beta\gamma^{-1}$
            $\alpha_j \leftarrow \alpha'_j$ for all $j \in 1, 2, \ldots, n'$.
        **end**
    **end**
  **V:** if $\beta = G_x(\alpha_1, \alpha_2, \ldots, \alpha_{n'})$ **then Halt** and **Accept**.
    **else Halt** and **Reject**.

Figure 5.1: A protocol for $L(M)$, for a restricted ATM $M$.

**Proof:**

1. This follows from the same arguments in the proof of Lemma 3.1.7.

2. This again follows from the similar arguments as in the proofs of Lemma 3.1.7 and Lemma 3.2.10. Note that in each elimination step the probability that the prover succeeds in that step is at most $\frac{5d}{p}$. In total, the probability that the verifier accepts the input is at most $\frac{5dN(n'+2)}{p} < \frac{1}{3}$.

3. The verifier uses $O(N((3+3+n'(|X|+1))T_{eval}+T_F)+T_{G_x}))$ additions and multiplications in $\mathbf{F}_p$, where $T_{eval}$ is the number of operations to evaluate a polynomial of degree $5d$ at one point, $T_F$ is the number of operations to evaluate $F$ at one point and $T_{G_x}$ is the number of operations to evaluate $G_x$ at one point.

   By Horner's Method $T_{eval} = O(5d)$. The time to evaluate the arithmetic formula of size $S$ is clearly $O(S)$. Hence the number of additions and multiplications performed by the verifier is

   $$O(t(n)s^{2+\epsilon}(n) + s^\epsilon(n)n \log n).$$

   Furthermore the verifier computes $O(t(n))$ inverses.

   The time for a Turing machine to perform one addition in $\mathbf{F}_p$ is $O(\log p)$ and multiplications can be done in time $O(\log^2 p)$. To compute an inverse using the extended GCD algorithm takes time $O(\log^3 p)$.

   Hence we infer that the verifier works in time

   $$O((t(n)s^{2+\epsilon}(n) + s^\epsilon(n)n \log n) \log^2 t(n)).$$

   The verifier uses space for two independent reasons. It is storing $O(n')$ field elements and it uses space to evaluate the polynomials and the formulas. Clearly a formula can be evaluated using only a number of registers proportional to the depth of the formula. Hence the verifier needs $O(n') = O(s(n)/\log s(n))$ registers each containing an element from $\mathbf{F}_p$. Hence the verifier uses

   $$O\left(\frac{s(n)\log p}{\log s(n)}\right) = O\left(\frac{s(n)\log t(n)}{\log s(n)}\right)$$

   space. ∎

This gives the main theorem of this chapter. Recall that $pIPTISP(t,s)$ is the class of languages for which there exists an interactive proof system with a public-coin verifier that works in time $O(t(n))$ and space $O(s(n))$.

**Theorem 5.3.2** *Given that* $(t,s)$ *is fully time-space constructible, if*

$$L \in ATISP(t(n), s(n))$$

*then L belongs to*

$$\bigcap_{\epsilon > 0} pIPTISP((s^2(n)t(n) + n\log n)s^\epsilon(n)\log^2 t(n), s(n)\log t(n)/\log s(n)).$$

**Corollary 5.3.3** *Every language in P has a public-coin interactive proof system with a polynomial time verifier using* $O(\log^2(n)/\log\log(n))$ *space.*

**Proof:** Let $L$ be a language in $P$. Chandra, Kozen and Stockmeyer [20] prove the existence of an alternating polynomial time log-space Turing machine $M$ that accepts $L$. ∎

**Corollary 5.3.4** *Every language in* $NC = \bigcup_k NC^k$ *has a public-coin interactive proof system with a verifier using* $O(\log(n))$ *space and* $O(n\log^2 n)$ *time.*

**Proof:** Ruzzo [63] shows that any language in $NC$ can be accepted by an alternating Turing machine using poly-log time and log space. Hence if $L \in NC$, then there exists a constant $k$ such that

$$
\begin{aligned}
L \;\in\;& ATISP(\log^k n, \log n) \\
\subseteq\;& \bigcap_{\epsilon > 0} pIPTISP(n(\log^{1+\epsilon} n)(\log\log n)^2, \log n) \\
\subseteq\;& pIPTISP(n\log^2 n, \log n). \;\blacksquare
\end{aligned}
$$

**Corollary 5.3.5** *If* $(t,s)$ *is fully time-space constructible, then*

$$ATISP(t(n), s(n)) \subseteq \bigcap_{\epsilon > 0} pIPTISP(n^{1+\epsilon} + t^{3+\epsilon}(n), s^2(n)/\log s(n)).$$

**Proof:** This follows from Theorem 5.3.2, since $t(n) \geq s(n)$ and $s(n) \geq \log t(n)$. ∎

# 5.4 Alternating Turing Machines for $pIPTISP(t, s)$

**Theorem 5.4.1** *Let $(t, s)$ be fully time-space constructible. Then*

$$pIPTISP(t(n), s(n)) \subseteq ATISP(t(n) \log t(n), s(n)).$$

**Proof:** Let $L$ have a public-coin interactive proof system using time $t(n)$ and space $s(n)$. We can assume without loss of generality that the protocol consists of exactly $t(n)$ rounds of the verifier sending a single coin toss to the prover followed by the prover sending back a one-bit response.

From any configuration $c$, the probability that the verifier accepts, starting in configuration $c$, must be $\frac{v_c}{2^{t(n)}}$ for some integer $v_c$, with $0 \leq v_c \leq 2^{t(n)}$. Let $v_c$ be the *value* of a configuration $c$. The value of an accepting configuration is 1 and the value of a rejecting configuration is 0.

If $c$ is the configuration immediately before a prover's message and if a prover response of 0 causes the verifier to enter configuration $c_0$ and a response of 1 causes the verifier to enter configuration $c_1$, then $v_c = \max(v_{c_0}, v_{c_1})$. If $c$ is the configuration immediately before a coin toss, $c_0$ is the configuration the verifier enters after tossing heads, and $c_1$ is the configuration the verifier enters after tossing tails, then $v_c = (v_{c_0} + v_{c_1})/2$.

An alternating machine to accept $L$ can work as follows: First existentially guess the value of the initial configuration and then verify its guess.

To verify its guess, we notice that maximum, addition (using carry lookahead) and division by 2 on $t(n)$-bit numbers has a uniform space $O(\log t(n))$ circuit of depth $O(\log t(n))$. So the value of the initial configuration can be calculated by a uniform (space $O(s(n))$) circuit of depth $O(t(n) \log t(n))$. By a result by Ruzzo [63] an alternating Turing machine can evaluate each bit of the value of the initial configuration in time $O(t(n) \log t(n))$ and space $O(s(n))$.

∎

Condon [21] and Fortnow and Sipser [34] independently proved the following fact, which follows from Theorem 5.4.1.

**Corollary 5.4.2** *A deterministic polynomial time Turing machine can recognize any language accepted by a public-coin interactive proof system with a verifier using logarithmic space and polynomial time.*

**Proof:** $\bigcup_{k>0} ATISP(n^k, \log n) = ASPACE(\log n) = P$ [20].  ∎

## 5.5   A Hierarchy for $pIPTISP(t, s)$

Theorem 5.1.2 gives a tight hierarchy for $ATISP(t, s)$.

**Theorem 5.5.1** *Given $(t_1, s_1)$ and $(t_2, s_2)$ fully time-space constructible pairs of functions,*

$$ATISP(t_1, s_1) \subsetneq ATISP(t_2, s_2)$$

*if*

$$t_1(n) = o(t_2(n)) \ and \ s_1(n) = o(s_2(n)).$$

**Proof:** Let $M_1, M_2, \ldots$ be an enumeration of 1-tape alternating Turing machines, such that every Turing machine has arbitrarily long encodings. We construct an alternating Turing machine $M$ that uses time $O(t_2(n))$ and space $O(s_2(n))$, and that recognizes a language not in $ATISP(t_1, s_1)$.

The idea is that $M$ tries to diagonalize against all the machines in the enumeration. It succeeds against all machines using time $O(t_1(n))$ and space $O(s_1(n))$.

On input $x$, $M$ simulates $M_x$ on input $x$, in the following way. If $M_x$ makes an existential guess then $M$ makes a universal guess and vice versa. Doing the simulation $M$ keeps track on the time and the space it is using. If on some computation path $M$ discover that it has used more than $t_2(n)$ time or $s_2(n)$ space and it halts and rejects the input. Otherwise the simulated machine $M_x$ halts and $M$ accepts if and only if $M_x$ rejects the input $x$. Note that if on all computation paths the simulation succeeds then $M$ accepts $x$ if and only if $M_x$ rejects.

Clearly $M$ uses $O(t_2(n))$ time and $O(s_2(n))$ space, since $(t_2, s_2)$ are time-space constructible. Hence $L(M) \in ATISP(t_2, s_2)$.

Assume that $L(M) \in ATISP(t_1, s_1)$. Hence we have an alternating Turing machine $M'$ that recognizes $L(M)$ and it works in time $ct_1(n)$ and space $cs_1(n)$, for some constant $c$. Note that given $M'$ there exists a constant $c'$ such that $M$ simulates $M'$ on an encoding of $M'$ with a slow down of at most $c'$ and uses only a factor of $c'$ more space than $M'$. Since there are arbitrarily long encodings of $M'$ let $x$ be an encoding of $M'$ such that

$$cc't_1(n) \leq t_2(n) \text{ and } cc's_1(n) \leq s_2(n).$$

Hence $M$ accepts $x$ if and only if $M_x$ rejects $x$. This contradict that $L(M) = L(M')$. ∎

We get a hierarchy theorem for $pIPTISP(t, s)$, because of our correspondence between $ATISP(t, s)$ and $pIPTISP(t, s)$.

**Theorem 5.5.2** *Given $(t_1, s_1)$ and $(t_2, s_2)$, fully time-space constructible pairs of functions such that $t_1, t_2 \geq n$ and $s_1, s_2 \geq \log n$,*

$$pIPTISP(t_1(n), s_1(n)) \subsetneq pIPTISP(t_2(n), s_2(n))$$

*if for some $\epsilon > 0$*

$$t_1^{3+\epsilon}(n) = o(t_2(n)) \text{ and } s_1^2(n) = o(s_2(n)).$$

**Proof:**

$$
\begin{aligned}
pIPTISP(t_1(n), s_1(n)) \quad &\subset \quad ATISP(t_1(n) \log t_1(n), s_1(n)) \\
&\subsetneq \quad ATISP(t_2(n)^{\frac{1}{3+\epsilon/2}}, s_2(n)^{\frac{1}{2}}) \\
&\subseteq \quad pIPTISP(n^{1+\epsilon/2} + t_2(n), s_2(n)) \\
&= \quad pIPTISP(t_2(n), s_2(n)).
\end{aligned}
$$

The first containment is from Theorem 5.4.1, the proper containment is Theorem 5.5.1 and the last containment is from Corollary 5.3.5. ∎

As a corollary we get that public-coin interactive proof systems with linear time verifiers can not recognize all of *IP*. This should be contrasted with the result by Fortnow and Sipser [36] that for probabilistic computation there exists an oracle $A$ such that $BPTIME(n)^A$ contains $BPP^A$. Furthermore, Theorem 5.5.2 gives a tighter hierarchy for time and space.

**Corollary 5.5.3** *For all reals* $1 \leq r < s$,

$$pIPTIME(n^r) \subsetneq pIPTIME(n^s),$$

$$pIPSPACE(n^r) \subsetneq pIPSPACE(n^s)$$

*and*

$$pIPSPACE(\log^r n) \subsetneq pIPSPACE(\log^s n).$$

**Proof:**    Given Theorem 5.5.2 the proof is similar to the proof of similar results for $NSPACE(s)$ by Ibarra [46] (see Theorem 12.12 in [45]).    ∎

# 5.6  Interactive Proof Systems for Deterministic Computation

Corollaries 5.3.3 and 5.3.4 exhibit interactive proof systems with verifiers having low time-space complexity for $P$ and $NC$. We can use Theorem 5.4.1 and Corollary 5.3.5 combined with the relationships in [20] described in Fact 2.3.3 to prove more general relationships.

**Corollary 5.6.1** *For* $t(n) \geq n, s(n) \geq \log n$,

- $\bigcup_{k>0} pIPTISP(t(n)^k, t(n)^k) = \bigcup_{k>0} DSPACE(t(n)^k)$.

- $\bigcup_{k>0} pIPTISP(2^{s(n)^k}, s(n)^k) = \bigcup_{k>0} DTIME(2^{s(n)^k})$.

*From this we get several consequences including:*

1. *An interactive protocol with a verifier using poly-log space and running in quasi-polynomial ($2^{\text{poly-log}(n)}$) time accepts the same set of languages as a deterministic Turing machine running in quasi-polynomial time.*

2. *A public-coin interactive protocol with a verifier running in polynomial time and space accepts exactly the same set of languages as a deterministic machine using polynomial space.*

3. *An interactive protocol with a verifier using polynomial space and exponential time accepts exactly the same set of languages as are deterministically recognizable in exponential time.*

4. *An interactive protocol with a verifier using exponential time and exponential space can accept all languages deterministically recognizable in exponential space.*

The second consequence is equivalent to *pIP = PSPACE*, which we proved in Chapter 3.

# 6

---

# Implications and Open Problems

The results in this book have implications in many different areas of theoretical computer science. In this chapter, we discuss these implications and areas for future research.

## 6.1 Bounded-Round Interactive Proofs

It has been an open problem whether every language in *IP* has a bounded-round interactive proof system. Previously, Aiello, Goldwasser and Hastad [1] constructed an oracle relative to which the class of languages with unbounded-round interactive proofs differs from those with bounded-round interactive proofs. Relative to any *PSPACE* -complete oracle, the answer is the opposite.

The following corollary gives strong evidence that not every language in *IP* has bounded-round interactive proofs.

**Corollary 6.1.1** *If every language in IP has a bounded round interactive proof, then* $PSPACE = \Pi_2^P$.

**Proof:** This is immediate from Babai [5], who shows that any language with a bounded-round interactive proof belongs to $\Pi_2^P$. ∎

# 6.2 Zero-Knowledge

The reason Goldwasser, Micali and Rackoff looked at interactive proofs was to study the amount of information that interactive protocols reveal. They introduced zero-knowledge interactive proof systems, which are proof systems where the prover only reveals the absolute minimum amount of information, *i.e.*, that $x \in L$.

In order for the prover to hide its information, we extend the model of interactive proof systems such that the prover can flip random coins that the verifier can not see.

To discuss zero-knowledge, we need some definitions. Given a interactive proof system $(P, V)$ that recognizes $L$, and an execution of the protocol, we define the *view* of the verifier to consist of the messages sent and the contents of its random tape. We think of the view as the information that the verifier has acquired during the execution of the protocol. Observe that since the prover is a probabilistic machine, the view depends on the contents of the random tape of the prover. By $(P, V)[x]$ we denote the probability distribution of views of the verifier on input $x$.

We say that the verifier did not get any information, except that $x \in L$, if there is a polynomial time probabilistic machine $M$, called the *simulator*, that can "simulate" the verifier's view of the protocol. We denote by $M[x]$ the probability distribution of $M$'s output on input $x$.

The case we want to guard against is that in which $x \in L$ and the honest prover is proving this to a verifier who is trying to acquire additional information. It may phrase its questions differently than in the original protocol, to trick the prover into revealing extra information.

Goldwasser, Micali and Rackoff [40] defined that a protocol $(P, V)$ does not reveal information if for all verifiers $V'$ there exists a simulator $M_{V'}$ such that for all $x \in L$, $M_{V'}$ can produce a distribution $M_{V'}[x]$ that is "close" to the distribution $(P, V')[x]$. There are a couple of definitions of "closeness."

- If $M_{V'}[x] = (P, V')[x]$ for all $x$ then we say that the protocol $(P, V)$ is *perfect zero-knowledge*.

- If $M_{V'}[x]$ and $(P, V')[x]$ are polynomial time indistinguishable then we say that the protocol $(P, V)$ is *zero-knowledge*, where $A[x]$ and $B[x]$ are polynomial time indistinguishable if for all polynomial time predicates $\varphi$, all polynomials $p$ and for all large $x$,

$$|\Pr[\varphi(A[x]) = 1] - \Pr[\varphi(B[x]) = 1]| < \frac{1}{p(|x|)}.$$

When they introduced multiple-prover interactive proof systems, Ben-Or, Goldwasser, Kilian and Wigderson, defined zero-knowledge and perfect zero-knowledge for multiple-prover interactive proofs in a similar fashion. They proved in [14] that every language in *MIP* has a perfect zero-knowledge multiple-prover interactive proof. Consequently, Corollary 6.2.1 follows from Theorem 4.3.4:

**Corollary 6.2.1** *Every language in NEXP has a perfect zero-knowledge multiple-prover interactive proof.*

A corresponding result for single-prover interactive proof systems would be surprising since Fortnow [30] proved that if a language $L$ has a perfect zero-knowledge single-prover interactive proof then $\overline{L}$ has a bounded-round interactive proof system and therefore Corollary 6.2.2 follows from Corollary 6.1.1 and Theorem 3.2.11:

**Corollary 6.2.2** *If every language in IP has a perfect zero-knowledge single-prover interactive proof then $PSPACE = \Pi_2^P$.*

On the other hand, Impagliazzo and Yung [48] and independently Ben-Or, Goldreich, Goldwasser, Hastad, Kilian, Micali and Rogaway [13], using a protocol by Goldwasser, Micali and Wigderson [39], showed that every language in *IP* has a zero-knowledge single-prover interactive proof given that honest one-way functions exists. A one-way function is a function $f$ that is computable in polynomial time, but for any polynomial size family of circuits $C_n$ and for every constant $c > 0$,

$$\Pr_{x \in \{0,1\}^n}[C_n(f(x)) \in f^{-1}(f(x))] < \frac{1}{n^c}$$

for all large enough $n$. A function $f$ is honest if there exists a constant $\epsilon > 0$ such that for all $x$, $|f(x)| \geq |x|^{\epsilon}$. Hence Corollary 6.2.3 follows from Theorem 3.2.11:

**Corollary 6.2.3** *If honest one way functions exist then every language in PSPACE has a zero-knowledge single-prover interactive proof.*

# 6.3 Program Testing and Verification

The results of this paper have many connections to program testing and verification. We make the connection precise in this section.

## 6.3.1 Robustness

In this section we describe a useful property of languages, *PSPACE*-robustness. We show that every *PSPACE*-robust language is Turing-equivalent to a family of multilinear functions (one $n$-variable function for every $n$).

**Definition 6.3.1** *A language $L$ is PSPACE-robust if $P^L = PSPACE^L$.*

Examples of *PSPACE*-robust languages include the *PSPACE*-complete and *EXP*-complete languages.

**Lemma 6.3.2** *Every language that is PSPACE-robust has a Turing-equivalent family of multilinear functions over the integers.*

**Proof:**  Let $L$ be a *PSPACE*-robust language. Let $g_n(x_1, \ldots, x_n)$ be the multilinear extension of the characteristic function of $L_n = L \cap \{0,1\}^n$ (see Proposition 2.1.1). Clearly $L \in P^g$, where $g = \{g_n : n \geq 0\}$. We describe an alternating polynomial time Turing machine $M$ with access to $L$ which computes $g$. First $M$ guesses the value $z = g_n(x_1, \ldots, x_n)$, then it existentially guesses the linear function $h_1(y) = g(y, x_2, \ldots, x_n)$ and verifies that $h_1(x_1) = z$.

Thereafter $M$ universally chooses $t_1 \in \{0, 1\}$ and existentially guesses the linear function $h_2(y) = g(t_1, y, x_3, \ldots, x_n)$. $M$ keeps repeating this process until $M$ has specified $t_1, \ldots, t_n$ and has a value $z$ for $g(t_1, \ldots, t_n)$. Then $M$ accepts if and only if

$$z = \begin{cases} 1 & \text{if } (t_1, \ldots, t_n) \in L, \\ 0 & \text{otherwise,} \end{cases}$$

which $M$ verifies using its oracle for $L$. Since a *PSPACE* machine can simulate an alternating polynomial time Turing machine, and since $L$ is *PSPACE*-robust, $g$ is Turing-reducible to $L$. ∎

In particular, we have multilinear *PSPACE*-complete functions, *EXP*-complete functions, etc. Inspired by Beaver and Feigenbaum [12] and spelled out simultaneously by us and by Beaver and Feigenbaum, this lemma has significant consequences, as we shall see below.

There are natural classes of languages satisfying the conclusion of Lemma 6.3.2 which are not known to be *PSPACE*-robust; $P^{\#P}$-complete languages being the prime example, since they are equivalent to the permanent, a multilinear function (Valiant [76]).

## 6.3.2 Instance Checking

In Blum and Kannan [16], *function-restricted IP* is defined as follows.

The set of all languages $L$ for which there is an interactive proof system for $L$ such that the honest prover need only answer questions about $L$, and any dishonest prover must be a function from the set of questions to $\{0, 1\}$. So in one way it is a restriction of *IP* since the honest prover only can answer questions about $L$, but in another hand it is a relaxation since the proof only has to guard against cheating provers that compute a function.

By Theorem 4.1.1 due to Fortnow, Rompel and Sipser we see that function-restricted *IP* is equivalent to multi-prover interactive proof systems where the honest provers can only answer questions about the language it is being asked to prove.

Blum and Kannan also define a *program checker* $C_L^{\mathcal{P}}$ for a language $L$ and an instance $x \in \{0,1\}^*$ as a probabilistic polynomial time oracle Turing machine that given a program $\mathcal{P}$ claiming to compute $L$, and an input $x$:

1. If $\mathcal{P}$ correctly computes $L$ for all inputs then with high probability $C_L^{\mathcal{P}}$ outputs "correct."

2. If $\mathcal{P}(x) \neq L(x)$, with high probability $C_L^{\mathcal{P}}(x)$ outputs "$\mathcal{P}$ does not compute $L$."

Blum and Kannan show that a language has a program checker if and only if the language and its complement each have a function-restricted interactive proof system.

The results in this book prove the following.

**Corollary 6.3.3** *Every $P^{\#P}$-complete language $L$ has a function-restricted interactive proof system and thus $L$ has a program checker.*

**Proof:** Theorem 3.1.9 shows that $L$ has a function-restricted interactive proof system and, since $P^{\#P}$ is closed under complement, $\overline{L}$ has one too, and therefore $L$ has a program checker. ∎

**Corollary 6.3.4** *Every $\#P$-complete function has a program checker.*

**Corollary 6.3.5** *Every PSPACE-complete language has a function-restricted interactive proof system and thus has a program checker.*

**Corollary 6.3.6** *Every EXP-complete language has a function-restricted interactive proof system and thus has a program checker.*

Thus not every language in function-restricted *IP* has a single-prover interactive proof unless *PSPACE = EXP*. This essentially gives a negative answer to the open question of Blum and Kannan [16] as to whether *IP* contains function-restricted *IP*.

Still open is the question whether *NP*-complete languages have program checkers. This is directly related to the question of whether *coNP* languages have protocols with *NP* provers.

## 6.3.3 Self-Testing and Self-Correcting Programs

Our test of multilinear functions (Section 4.4) also has applications to program testing as described by Blum, Luby and Rubinfeld [17] and Lipton [54].

We use the following definition of self-testing/correcting programs slightly different from but in the spirit of the Blum-Luby-Rubinfeld definition. We make the connection between the two models clear in Section 6.3.4.

An input set $I$ is a sequence of subsets $I_1, I_2, \ldots$ of $\{0,1\}^*$ such that for some $k$ and for all $n$, if $x \in I_n$ then $n^{1/k} \leq |x| \leq n^k$. We let $I$ represent the set $\cup_{n \geq 1} I_n$.

We say a pair of probabilistic polynomial time programs $(T, C)$ is a *self-testing/correcting pair* for a function $f$ over an input set $I$ if given a program $\mathcal{P}$ that purports to compute $f$ the following hold for every $n$:

1. The tester $T(\mathcal{P}, 1^n)$ outputs either "Pass" or "Fail."

2. If $\mathcal{P}$ correctly computes $f$ on all inputs of $I$ then $T(\mathcal{P}, 1^n)$ says "Pass" with probability at least $2/3$.

3. For every $x \in I_n$, if $\Pr(T(\mathcal{P}, 1^n)$ says "Pass"$) > 1/3$ then $\Pr(C(\mathcal{P}, x) = f(x)) > 2/3$.

The errors can be made exponentially small by repeated trials and majority vote. A language has a self-testing/correcting pair if its characteristic function does.

An alternative definition would require the tester to say always "Pass" for a correct program. In every case that we know, the tester has this property. However, we allow the more general definition for a better comparison with the Blum-Luby-Rubinfeld model (see Section 6.3.4).

**Theorem 6.3.7** *Every PSPACE-complete and EXP-complete language has a self-testing/correcting pair over $I = \{I_n | I_n = \{0,1\}^n\}$.*

We will prove Theorem 6.3.7 for EXP-complete languages. The proof for PSPACE-complete languages is analogous. In fact this proof holds for any language $L$ that is PSPACE-robust and has a multiple-prover interactive proof system where the provers only answer questions about membership in $L$.

**Lemma 6.3.8** *Let $g_n$ be the multilinear extension of an $EXP$-complete language $L$ over the field of $p_n$ elements where $p_n$ is the least prime greater than $n$. The function $g = \{g_n : n \geq 0\}$ has a self-testing/correcting pair over the set $I = \{I_n | I_n = \mathbf{F}_{p_n}^n\}$.*

**Proof:** Since $g$ is $EXP$-hard and each bit of $g_n(x_1, \ldots, x_n)$ is computable in $EXP$ (Lemma 6.3.2), by Corollary 4.3.5 there exists a multiple-prover interactive proof system for verifying a specific bit of $g_n(x_1, \ldots, x_n)$ where the provers need only answer questions about $g$.

Let $\mathcal{P}$ be a program that purports to compute $g$. The tester program $T(\mathcal{P}, 1^n)$ chooses $n^3$ randomly chosen $(y_1, \ldots, y_n) \in \mathbf{F}_{p_n}^n$. $T$ then verifies that each bit of the $\mathcal{P}(y_1, \ldots, y_n)$ is the same as $g(y_1, \ldots, y_n)$ with a multiprover interactive proof system using $\mathcal{P}$ as the provers. The tester $T$ outputs "Pass" if every bit checks correctly. If $T$ outputs "Pass" with probability at least $1/3$ then with extremely high confidence $\mathcal{P}(y_1, \ldots, y_n) = g_n(y_1, \ldots, y_n)$ on all but $1/n^2$ of the possible $(y_1, \ldots, y_n) \in \mathbf{F}_{p_n}^n$.

We use ideas of Beaver-Feigenbaum [12] and Lipton [54] to create the correcting function $C$. Suppose we wish to compute $g_n(x_1, \ldots, x_n)$. Choose elements $r_1, \ldots, r_n \in \mathbf{F}_{p_n}^n$ at random and let $r^i = (x_1 + ir_1, \ldots, x_n + ir_n)$ for $1 \leq i \leq n+1$. Let $g'(y) = g_n(x_1 + yr_1, \ldots, x_n + yr_n)$ for all $y$. With probability greater than $1 - \frac{p_n}{n^2}$ (By "Bertrand's Postulate", $p_n < 2n$), $\mathcal{P}(r^i) = g_n(r^i) = g'(i)$ since each $r^i$ is uniformly random. However, $g'(y)$ is a polynomial of degree at most $n$ and we have $n+1$ points of this polynomial, $g'(1), \ldots, g'(n+1)$. Interpolate this polynomial and compute $g'(0) = g_n(x_1, \ldots, x_n)$. If we repeat this process $n$ times then with extremely high probability a majority of the answers from this process is the proper value of $g_n$. ∎

We can now prove Theorem 6.3.7:

**Proof:**    Suppose we had a program $\mathcal{Q}$ that purports to compute $L$. By Lemma 6.3.2 there exists a polynomial time function $f(y_1, \ldots, y_n, i)$ such that the $i$th bit of $g_n(y_1, \ldots, y_n)$ is one if and only if $f(y_1, \ldots, y_n, i) \in L$. We create a new program $\mathcal{P}$ that simulates this process asking questions to $\mathcal{Q}$ instead of $L$. If $\mathcal{Q}$ properly computes $L$ then $\mathcal{P}$ properly computes $g_n$.

Let $T_g$ and $C_g$ be the testing/self-correcting pair for $g$. We will create a testing/self-correcting pair $T_L, C_L$ for $L$. The tester $T_L(\mathcal{Q}, 1^n)$ just simulates $T_g(\mathcal{P}, 1^n)$ using the $\mathcal{P}$ described above. The checker $C_L(\mathcal{Q}, x)$ just outputs $C_g(\mathcal{P}, (x_1, \ldots, x_n))$ where $x = x_1 \ldots x_n$.  ∎

## 6.3.4    Comparison with the Blum-Luby-Rubinfeld Model

Blum, Luby and Rubinfeld [17] give the following series of definitions for self-testing/correcting pairs:

Let $\mathcal{D} = \{\mathcal{D}_n | n \geq 0\}$ be an ensemble of probability distributions such that $\mathcal{D}_n$ is a distribution on $I_n$. Let $\mathcal{P}$ be a program that purports to compute $g$. Let $\text{error}(g, \mathcal{P}, \mathcal{D}_n)$ be the probability that $\mathcal{P}(x) \neq g(x)$ when $x$ is chosen from $\mathcal{D}_n$.

Let $0 \leq \epsilon_1 < \epsilon_2 \leq 1$. An $(\epsilon_1, \epsilon_2)$-*self-testing* program for $g$ with respect to $\mathcal{D}$ is a probabilistic polynomial time program $T$ such that

1. If $\text{error}(g, \mathcal{P}, \mathcal{D}_n) \leq \epsilon_1$ then $T(\mathcal{P}, 1^n)$ outputs "Pass" with probability at least 2/3.

2. If $\text{error}(g, \mathcal{P}, \mathcal{D}_n) \geq \epsilon_2$ then $T(\mathcal{P}, 1^n)$ outputs "Pass" with probability at most 1/3.

Let $0 \leq \epsilon < 1$. An $\epsilon$-*self-correcting* program for $f$ with respect to $\mathcal{D}$ is a probabilistic polynomial time program $C$ such that for all $x \in I_n$, if $\text{error}(g, \mathcal{P}, \mathcal{D}_n) \leq \epsilon$ then $C(\mathcal{P}, x) = g(x)$ with probability at least 2/3.

A *self-testing/correcting* pair for $g$ over an input set $I$ is a pair of programs $(T, C)$ such that for some $\epsilon, \epsilon_1, \epsilon_2$ with $0 \leq \epsilon_1 < \epsilon_2 \leq \epsilon < 1$ and some ensemble of probability distributions $\mathcal{D}$ over $I$ such that $T$ is a $(\epsilon_1, \epsilon_2)$-self-testing program for $g$ with respect to $\mathcal{D}$ and $C$ is an $\epsilon$-self-correction for $g$ with respect to $\mathcal{D}$.

Note that if $g$ has a self-testing/correcting pair $(T, C)$ over an input set $I$ in the Blum-Luby-Rubinfeld model then $g$ has a self-testing/correcting pair in our model using the same $T$ and $C$.

**Lemma 6.3.9** *If $L$ is PSPACE-robust and has a function-restricted interactive proof system then there exists a family $g$ of multilinear functions Turing-equivalent to $L$ that has a self-testing/correcting pair using the Blum-Luby--Rubinfeld model.*

**Proof:**   Use the function $g$ defined in Lemma 6.3.8. The same tester and corrector $T$ and $C$ used in the proof of Lemma 6.3.8 also work here. Let $\mathcal{D}_n$ be the uniform distribution over $\mathbf{F}_{p_n}^n$. The tester $T$ is a $(0, 1 - 1/n^2)$-self-testing program for $g$ over $\mathcal{D}$. The corrector $C$ is a $1 - 1/n^2$-self-correcting program for $g$ over $\mathcal{D}$.  ∎

**Corollary 6.3.10** *There exist PSPACE-complete and EXP-complete functions that have self-testing/correcting pairs in the Blum-Luby-Rubinfeld model.*

It's not clear whether all PSPACE-complete or EXP-complete languages have self-testing/correcting pairs under the Blum-Luby-Rubinfeld model.

We can also do program verification in the spirit of Lipton (Blum-Luby-Rubinfeld without an assumption of a tester $T$) as follows: Suppose a program $\mathcal{P}$ claims to compute a multilinear function. We can test that there is some multilinear function $f$ using Theorem 4.4.1 such that $\mathcal{P} = f$ on most inputs and then if $\mathcal{P} = f$ on most inputs we can create a correcting function $C$ such that $C = f$ on all inputs with high probability. The proof is virtually identical to the proof of Lemma 6.3.8. We can also replace "multilinear" by "small-degree polynomial" as defined in Theorem 4.4.17.

# 6.4   Uniform vs. Nonuniform Complexity

Karp and Lipton [49] have considered the effect of nonuniform simulation of
large complexity classes by small circuits on uniform complexity classes. They
credit A. Meyer for one of following results ($\mathcal{C} = EXP$):

**Theorem 6.4.1 (Meyer, Karp, Lipton)** *Let $\mathcal{C}$ be one of the following complexity classes: $EXP$, $PSPACE$, $P^{\#P}$. If $\mathcal{C}$ has polynomial size circuits (i.e.,
$\mathcal{C} \subset P/poly$) then $\mathcal{C} = \Sigma_2^P$.*

The results described in this book lead to a strengthening of the conclusion
in each case, replacing $\Sigma_2^P$ by its subclass $MA$.

**Corollary 6.4.2** *If every language $L \in \mathcal{C}_1$ has a multiple-prover interactive
proof system with provers that live in $\mathcal{C}_2$ and if $\mathcal{C}_2$ has polynomial-size circuits,
then $L \in MA$.*

**Proof:**   The prover produces the circuits for $L_1, \ldots, L_k$ that describe the
responses for provers $P_1, \ldots, P_k$, respectively, and the verifier simulates the
verifier for $L$ using the circuits to compute the provers' responses.   ∎

In particular, if $L$ has a function-restricted interactive proof and $L$ has
polynomial-size circuits, then $L \in MA$.

**Corollary 6.4.3** *If every language in $EXP$ has polynomial-size circuits then
$EXP = \Sigma_2^P = \Pi_2^P = MA$. The same statement holds for $PSPACE$ and $P^{\#P}$ in
place of $EXP$.*

It seems remarkable that this result, which refers to standard concepts of
structural complexity theory, has been proved via the theory of multi-prover
interactive proof systems.

# 6.5   Recent Developments

Since the discovery of the results described in this book, researchers have build on these ideas and come up with new proof systems and new application of interactive proofs.

The multiple-prover interactive proof system from Chapter 4 has been scaled down to prove efficiently *NP*-statements.

Babai, Fortnow, Levin and Szegedy [8] invented the concept of *transparent proofs* and they showed that every *NP*-statement has such a proof, *i.e.*, every formal mathematical proof $P$ can be transformed into a transparent proof $P'$, that can be verified in probabilistic time polynomial in the logarithm for the size of $P$ and furthermore, that the size of $P'$ is polynomially related to the size of $P$.

Feige, Goldwasser, Lovász, Safra and Szegedy [26] discovered that the multiple-prover interactive proof system results implies that clique-approximation is hard. They showed that if some polynomial time procedure can, given a graph, approximate the size of the maximal clique (MAX-CLIQUE) of the graph within a constant factor then $NP \subset DTIME(n^{\log\log n})$. In order to obtain this result they constructed an oracle proof system for 3-SAT such that the verifier only uses $O(\log n \log\log n)$ random bits and only accesses $O(\log n \log\log n)$ bits from the oracle.

Arora and Safra [4] improved the clique result. They showed that a constant fraction approximation procedure for MAX-CLIQUE would imply that $P = NP$. They did so by applying the transparent proof idea recursively to obtain an oracle proof system for 3-SAT such that the verifier only uses $O(\log n)$ random bits and only accesses $o(\log n)$ bits from the oracle.

Arora, Lund, Motwani, Sudan and Szegedy [3] have recently constructed an even more efficient oracle proof system for 3SAT where the verifier only uses $O(\log n)$ random bits and only accesses a constant number of bits from the oracle. This improved the MAX-CLIQUE results even further. More importantly, they showed that their oracle proof system implies that MAX-SNP-hard problems [59] does not have a polynomial time approximation scheme unless $P = NP$. For example, there exists a constant $\epsilon > 0$ such that if given a 3-CNF formula then computing an assignment of the variable, such that the number of satisfied clause are within a factor of $(1 - \epsilon)$ from an optimal assignment is *NP*-hard.

Lund and Yannakakis [56] have very recently shown other hardness results for approximation problems using interactive proof systems results. They showed hardness results for coloring, setcover and a wide variety of maximal subgraph problems.

A lot of work went into parallelizing the multiple-prover protocol from Chapter 4. This was obtained by a series of result [19, 51, 25, 52, 27], which ended up by proving that every language in *NEXP* has a two-prover one-round interactive proof system that has exponentially small error.

The ideas in Chapter 3 have been used by Feige and Lund [28] to show that the permanent is hard on almost all inputs unless the polynomial time hierarchy collapses.

## 6.6   Further Research

Many interesting open questions remain, including

- Which languages $L$ can be proven by bounded-round interactive proofs? From Babai and Moran [10] we know that if $L$ has a bounded-round interactive proof, then $L \in \Pi_2^P$. But, is such an $L$ in $\Sigma_2^P$?

- What complexity of provers do we need to prove *coNP* and *NEXP* languages? Is there a difference for *coNP* when we only have one prover?

- Can the time-space relationship in Chapter 5 be tightened? In particular, can we show that the class of languages accepted by an interactive proof system with a polynomial time, log-space verifier is equal to P?

- Can we prove similar kinds of relationships as in Chapter 5 for different models? For example, consider the multiple-person alternation model of Peterson and Reif [61] and the multiple-prover interactive proof system.

The work in Chapter 4 relating multi-prover protocols to nondeterministic exponential time suggests a time-space relationship between the multiple-person alternation model and the multiple-prover interactive proof system model.

- Finally, this seems an occasion to express a cautious hope that the techniques discussed above might lead to a solution of some long-standing separation problems such as *BPP* vs. *NEXP* (cf. [43], [44]). Although there exist oracles which collapse these classes, this fact no longer seems as discouraging as it used to be, in view of a substantial mass of new techniques that do not relativize.

# Bibliography

[1] W. Aiello, S. Goldwasser, and J. Hastad. On the power of interaction. *Combinatorica*, 10(1):3–25, 1990.

[2] D. Aldous. On the Markov chain simulation method for uniform combinatorial distributions and simulated annealing. *Probability in the Engineering and Informational Sciences*, 1:33–46, 1987.

[3] S. Arora, C. Lund, R. Motwani, M. Sudan, and M. Szegedy. Proof verification and intractability of approximation problems, 1992. Manuscript.

[4] S. Arora and S. Safra. Approximating clique is NP-complete, 1992. Manuscript.

[5] L. Babai. Trading group theory for randomness. In *Proc. of the 17th ACM Symp. on the Theory of Computing*, pages 421–429, 1985.

[6] L. Babai and P. Erdős. Representation of group elements as short products. *Annals of Discrete Mathematics*, 12:27–30, 1982.

[7] L. Babai and L. Fortnow. Arithmetization: a new method in structural complexity theory. *Computational Complexity*, 1:41–66, 1991.

[8] L. Babai, L. Fortnow, L. Levin, and M. Szegedy. Checking computations in polylogarithmic time. In *Proc. of the 23rd ACM Symp. on the Theory of Computing*, pages 21–31, 1991.

[9] L. Babai, L. Fortnow, and C. Lund. Non-deterministic exponential time has two-prover interactive protocols. *Computational Complexity*, 1:3–40, 1991.

[10] L. Babai and S. Moran. Arthur-Merlin games: a randomized proof system, and a hierarchy of complexity classes. *J. of Computer and System Sciences*, 36(2):254–276, 1988.

[11] T. Baker, J. Gill, and R. Solovay. Relativizations of the P = NP question. *SIAM J. of Computing*, 4(4):431–442, 1975.

[12] D. Beaver and J. Feigenbaum. Hiding instances in multioracle queries. In *Proc. 7th Symp. on Theoretical Aspects of Comp. Sci.*, pages 37–48. LNCS 415, 1990.

[13] M. Ben-Or, O. Goldreich, S. Goldwasser, J. Hastad, J. Kilian, S. Micali, and P. Rogaway. Everything provable is provable in zero-knowledge. In *Proc. Crypto 88*, pages 37–56, 1988.

[14] M. Ben-Or, S. Goldwasser, J. Kilian, and A. Wigderson. Multi-prover interactive proofs: How to remove intractability assumptions. In *Proc. of the 20th ACM Symp. on the Theory of Computing*, pages 113–131, 1988.

[15] C. Bennet and J. Gill. Relative to a random oracle, $P^A \neq NP^A \neq coNP^A$ with probability one. *SIAM J. on Computing*, 10:96–113, 1981.

[16] M. Blum and S. Kannan. Designing programs that check their work. In *Proc. of the 21st ACM Symp. on the Theory of Computing*, pages 86–97, 1989.

[17] M. Blum, M. Luby, and R. Rubinfeld. Self-testing and self-correcting programs, with applications to numerical programs. In *Proc. of the 22nd ACM Symp. on the Theory of Computing*, 1990.

[18] A. R. Bruss and A. R. Meyer. On time-space classes and their relation to the theory of real addition. *Theoretical Computer Science*, 11:59–69, 1980.

[19] J. Cai, A. Condon, and R. J. Lipton. PSPACE is provable by two provers in one round. In *Proc. of the 6th Conference on Structure in Complexity Theory*, pages 110–115, 1991.

[20] A. Chandra, D. Kozen, and L. Stockmeyer. Alternation. *J. of the ACM*, 28(1):114–133, 1981.

[21] A. Condon. Space bounded probabilistic game automata. In *Proc. of the 3rd Conference on Structure in Complexity Theory*, pages 162–174, 1988.

[22] A. Condon and R. J. Lipton. On the complexity of space bounded interactive proofs. In *Proc. of the 30th IEEE Symp. on Foundations of Computer Science*, pages 462–467, 1989.

[23] S. Cook. The complexity of theorem-proving procedures. In *Proc. of the 3rd ACM Symp. on the Theory of Computing*, pages 151–158, 1971.

[24] S. Cook. Determinstic CFLs are accepted simultaneous in polynomial time and log squared space. In *Proc. of the 11st ACM Symp. on the Theory of Computing*, pages 338–345, 1979.

[25] U. Feige. On the success probability of the two provers in one round proof systems. In *Proc. of the 6th Conference on Structure in Complexity Theory*, pages 116–123, 1991.

[26] U. Feige, S. Goldwasser, L. Lovász, S. Safra, and M. Szegedy. Approximating clique is almost NP-complete. In *Proc. of the 32th IEEE Symp. on Foundations of Computer Science*, pages 2–12, 1991.

[27] U. Feige and L. Lovasz. Two-prover one-round proof systems: Their power and their problems. In *Proc. of the 24th ACM Symp. on the Theory of Computing*, 1992. To appear.

[28] U. Feige and C. Lund. On the hardness of computing the permanent of random matrices. In *Proc. of the 24th ACM Symp. on the Theory of Computing*, 1992. To appear.

[29] P. Feldman. The optimum prover lives in *PSPACE*. Manuscript, M.I.T., 1986.

[30] L. Fortnow. The complexity of perfect zero-knowledge. In S. Micali, editor, *Randomness and Computation*, volume 5 of *Advances in Computing Research*, pages 327–343. JAI Press, 1989.

[31] L. Fortnow. *Complexity-theoretic aspects of interactive proof systems*. PhD thesis, Massachusetts Institute of Technology, Laboratory for Computer Science, 1989. Tech Report MIT/LCS/TR-447.

[32] L. Fortnow and C. Lund. Interactive proof systems and alternating time-space complexity. In *Proc. 8th Symp. on Theoretical Aspects of Comp. Sci.*, pages 263–274. LNCS 480, 1991. To appear in Theoretical Computer Science.

[33] L. Fortnow, J. Rompel, and M. Sipser. On the power of multi-prover interactive protocols. In *Proc. of the 3rd Conference on Structure in Complexity Theory*, pages 156–161, 1988.

[34] L. Fortnow and M. Sipser. Interactive proof systems with a log-space verifier. Manuscript. Later version appears in [31].

[35] L. Fortnow and M. Sipser. Are there interactive protocols for co-NP languages? *Information Processing Letters*, 28:249–251, 1988.

[36] L. Fortnow and M. Sipser. Probabilistic computation and linear time. In *Proc. of the 21st ACM Symp. on the Theory of Computing*, pages 148–156, 1989.

[37] M. R. Garey and D. S. Johnson. *Computers and intractability. A Guide to the theory of NP-completeness*. W. H. Freeman and Company, New York, 1979.

[38] O. Goldreich, Y. Mansour, and M. Sipser. Interactive proof systems: Provers that never fail and random selection. In *Proc. of the 28th IEEE Symp. on Foundations of Computer Science*, pages 449–461, 1987.

[39] O. Goldreich, S. Micali, and A. Wigderson. Proofs that yield nothing but their validity and a methodology of cryptographic protocol design. In *Proc. of the 27th IEEE Symp. on Foundations of Computer Science*, pages 174–187, 1986.

[40] S. Goldwasser, S. Micali, and C. Rackoff. The knowledge complexity of interactive proof-systems. *SIAM J. on Computing*, 18(1):186–208, 1989.

[41] S. Goldwasser and M. Sipser. Private coins versus public coins in interactive proof systems. In S. Micali, editor, *Randomness and Computation*, volume 5 of *Advances in Computing Research*, pages 73–90. JAI Press, 1989.

[42] G. H. Hardy and E. M. Wright. *An introduction to the theory of numbers*. Oxford University Press, New York, New York, 5 edition, 1974.

[43] J. Hartmanis, N. Immerman, and V. Sewelson. Sparse sets in $NP - P$ : *EXPTIME* versus *NEXPTIME*. *Inf. and Control*, 65:158–181, 1985.

[44] H. Heller. On relativized exponential and probabilistic complexity classes. *Information and Computation*, 71:231–243, 1986.

[45] J. E. Hopcroft and J. D. Ullman. *Introduction to Automata Theory, Languages and Computation*. Addison-Wesley, Reading, Mass., 1979.

[46] O. H. Ibarra. A note concerning nondeterministic tape complexities. *J. ACM*, 19(4):608–612, 1972.

[47] N. Immerman. Nondeterministic space is closed under complement. In *Proc. of the 3rd Conference on Structure in Complexity Theory*, pages 112–115, 1988.

[48] R. Impagliazzo and M. Yung. Direct minium-knowledge computation. In *Proc. Crypto 87*, pages 40–51. LNCS 293, 1987.

[49] R. Karp and R. Lipton. Some connections between nonuniform and uniform complexity classes. In *Proc. of the 12th ACM Symp. on the Theory of Computing*, pages 302–309, 1980.

[50] R. M. Karp. Reducibility among combinatorial problems. In R. E. Miller and J. W. Thatcher, editors, *Complexity of Computer Computations*, Advances in Computing Research, pages 85–103. Plenum Press, 1972.

[51] J. Kilian. Interactive proofs based on the speed of light, 1991. e-mail announcement.

[52] D. Lapidot and A. Shamir. Fully parallelized multi prover protocols for NEXP-time. In *Proc. of the 32th IEEE Symp. on Foundations of Computer Science*, pages 13–18, 1991.

[53] L. Levin. Universal'nyĭe perebornyĭe zadachi (Universal search problems : in Russian). *Problemy Peredachi Informatsii*, 9(3):265–266, 1973.

[54] R. Lipton. New directions in testing. In J. Feigenbaum and M. Merritt, editors, *Distributed Computing and Cryptography*, volume 2 of *DIMACS Series in Discrete Mathematics and Theoretical Computer Science*, pages 191–202. American Mathematical Society, 1991.

[55] C. Lund, L. Fortnow, H. Karloff, and N. Nisan. Algebraic methods for interactive proof systems. In *Proc. of the 31th IEEE Symp. on Foundations of Computer Science*, pages 2–10, 1990. To appear in J. of the ACM.

[56] C. Lund and M. Yannakakis, 1992. Manuscript.

[57] P. Orponen. Complexity classes of alternating machines with oracles. In *Proc. of the 10th International Colloquium on Automata, Languages and Programming*, pages 573–584. LNCS 154, 1983.

[58] C. Papadimitriou. Games against nature. *J. of Computer and System Sciences*, 31:288–301, 1985.

[59] C. Papadimitriou and M Yannakakis. Optimization, approximation and complexity classes. In *Proc. of the 20th ACM Symp. on the Theory of Computing*, pages 229–234, 1988.

[60] W. J. Paul, E. J. Prauß, and R. Reischuk. On alternation. *Acta Informatica*, 14:243–255, 1980.

[61] G. Peterson and J. Reif. Multiple-person alternation. In *Proc. of the 20th IEEE Symp. on Foundations of Computer Science*, pages 348–363, 1979.

[62] V. Pratt. Every prime has a succinct certificate. *SIAM J. Comp*, 4:214–220, 1975.

[63] W. Ruzzo. On uniform circuit complexity. *J. of Computer and System Sciences*, 22:365–381, 1981.

[64] W. J. Savitch. Relationship between nondeterministic and deterministic tape complexities. *J. Computer System Science*, 4:177–192, 1970.

[65] J. T. Schwartz. Probabilistic algorithms for verification of polynomial identities. *J. of the ACM*, 27:701–717, 1980.

[66] A. Shamir. *IP =PSPACE* . In *Proc. of the 31st IEEE Symp. on Foundations of Computer Science*, pages 11–15, 1990. To appear in J. of the ACM.

[67] A. Shen. *IP = PSPACE*: simplified proof. Manuscript, 1990.

[68] J. Simon. *On some central problems in computational complexity.* PhD thesis, Cornell University, Computer Science, 1975. Tech Report TR 75-224.

[69] L. J. Stockmeyer. The polynomial-time hierachy. *Theoretical Computer Science*, 3:1–22, 1976.

[70] L. J. Stockmeyer and A. R. Meyer. Word problems requiring exponential time. In *Proc. of the 5th ACM Symp. on the Theory of Computing*, pages 1–9, 1973.

[71] I. Sudborough. Time and tape bounded auxiliary pushdown automata. *Mathematical Foundations of Computer Science*, pages 493–503, 1977.

[72] I. Sudborough. On the tape complexity of deterministic context free languages. *J. of the ACM*, 25(3):405–414, 1978.

[73] R. Szelepcsényi. The method of forcing for nondeterministic automata. *Bull. European Association Theoretical Computer Science*, pages 145–152, 1987.

[74] S. Toda. On the computational power of PP and $\oplus$P. In *Proc. of the 30th IEEE Symp. on Foundations of Computer Science*, pages 514–519, 1989.

[75] A. Turing. On computable numbers with an application to the Entscheidungs-problem. *Proc. London Math. Soc.*, 42(2):230–265, 1936. A correction, *ibid.*, 43:544–546.

[76] L. Valiant. The complexity of computing the permanent. *Theoretical Computer Science*, 8:189–201, 1979.

# Index

The MIT Press, with Peter Denning as general consulting editor, publishes computer science books in the following series:

**ACL-MIT Press Series in Natural Language Processing**
Aravind K. Joshi, Karen Sparck Jones, and Mark Y. Liberman, editors

**ACM Doctoral Dissertation Award and Distinguished Dissertation Series**

**Artificial Intelligence**
Patrick Winston, founding editor
J. Michael Brady, Daniel G. Bobrow, and Randall Davis, editors

**Charles Babbage Institute Reprint Series for the History of Computing**
Martin Campbell-Kelly, editor

**Computer Systems**
Herb Schwetman, editor

**Explorations with Logo**
E. Paul Goldenberg, editor

**Foundations of Computing**
Michael Garey and Albert Meyer, editors

**History of Computing**
I. Bernard Cohen and William Aspray, editors

**Logic Programming**
Ehud Shapiro, editor; Fernando Pereira, Koichi Furukawa, Jean-Louis Lassez, and David H. D. Warren, associate editors

**The MIT Press Electrical Engineering and Computer Science Series**

**Research Monographs in Parallel and Distributed Processing**
Christopher Jesshope and David Klappholz, editors

**Scientific and Engineering Computation**
Janusz Kowalik, editor

**Technical Communication and Information Systems**
Edward Barrett, editor